T0222442

.

Undergraduate Texts in Mathematics

# Undergraduate Texts in Mathematics

**Undergraduate Texts in Mathematics** are generally aimed at third- and fourth-year undergraduate mathematics students at North American universities. These texts strive to provide students and teachers with new perspectives and novel approaches. The books include motivation that guides the reader to an appreciation of interrelations among different aspects of the subject. They feature examples that illustrate key concepts as well as exercises that strengthen understanding.

More information about this series at http://www.springer.com/series/666

Donald G. Saari

# Mathematics of Finance

## An Intuitive Introduction

 Springer

Donald G. Saari
Departments of Economics
and Mathematics
University of California
Irvine, CA, USA

ISSN 0172-6056          ISSN 2197-5604   (electronic)
Undergraduate Texts in Mathematics
ISBN 978-3-030-25442-1          ISBN 978-3-030-25443-8   (eBook)
https://doi.org/10.1007/978-3-030-25443-8

Mathematics Subject Classification: 91Gxx, 91G20, 91F99

This Springer imprint is published by the registered company Springer Nature Switzerland AG.
The registered company address is: Gewerbestrasse 11, 6330 Cham, Switzerland

*For my good friend*

*Arthur Pancoe*

*A true financial wizard!*

# Note to Instructors

These notes were motivated by conversations with graduating seniors in mathematics and economics who were headed for the finance world. Finance? From where did they get their training? It turned out at that time at Northwestern University there were no relevant undergraduate courses. And so, after discussions with colleagues from the Departments of Finance and of Economics, I developed a course for the Mathematics Department.

Based on the advantages gained by students who have taken this course, including job opportunities and a new focus on course material, I believe that such a course should be offered in most undergraduate math programs. And so, this book is designed to help the reader grasp the fundamentals. This is both for students and for instructors who wish to teach the course but may be hesitant without a previous background. Course enrollments have been a healthy mixture where about half are math majors and the other half are mathematically stronger students from economics and elsewhere (e.g., engineering, graduate students, etc.). After my move to the sunnier climes at the University of California, Irvine, the notes were modified to meet the realities of larger class sizes.

The course addresses several needs:

1. *Capstone:* Students learn a considerable amount of mathematics as undergraduates, but many fail to see how it is connected, what to do with it, or how it is relevant for their future. Even more, many either forgot fundamental concepts or never moved beyond technical details to grasp the absorbing power of mathematics.

   As an example, it is not unusual to find students who dismiss Taylor series as a side topic or as an illustration of infinite series. They most surely carried out numerous exercises yet failed to recall that, for practical purposes, this powerful tool requires only a finite number of terms. Then, many fail to remember how to create Taylor series for several variables. The material was taught, but for some (not all), it was forgotten.

   So, rather than *assuming* students recall material from earlier courses (including basic concepts from probability and statistics), the relevant concerns are quickly reviewed with an emphasis on their power and utility. This is *not* a course on these subjects, so an intuitive review, rather than a detailed, rigorous exposition, is presented. A nice feature of this finance topic is that it incorporates so many mathematical concepts, which explain why the course has served as a capstone for students with only a passing curiosity about finance.

2. *Introduction:* The main purpose is to introduce students to the fundamentals of the mathematics of finance. Most arrive knowing nothing about this area, so this book starts with basics and quickly moves to more complicated material. The choice of material is directed to provide a mathematical understanding of the fundamentals with an emphasis on *why* certain equations and concepts are of value and what they really mean.

   It is standard in some courses, for instance, to present the solution for the Black–Scholes Equation without explaining from where terms come, why they should be expected, and what they mean. One way to close this gap is to carefully solve the Black–Scholes Equation. But once the course's popularity attracted classes of around 200 students, time limitations made this impossible.

   Fortunately, the Black–Scholes solution can be understood without solving the partial differential equation; it suffices to appreciate how changes of variables (to convert the Black–Scholes Equation into the heat equation) are manifested in the final solution. In this manner, students understand from where and why terms arise rather than confronting a confusing memory lesson. Attention can then be focussed on what all of this means.

3. *Developing mathematical intuition:* Students who have taken this course have directly entered the finance world or moved to graduate programs to learn more. This means that they must develop intuition for what is being presented, the limitations of various conclusions,[1] and what topics are open for research.

   Limitations are emphasized throughout the book starting with the introduction. Students catch on: they begin to appreciate the importance of those hypotheses that, in the past, they might have ignored. Better students recognize where added research is required.

   To help students develop an instinctive understanding of the material, the approach of this book differs from a traditional course: new material is introduced in terms of what they can readily grasp. That is, topics are launched with stories or closely related themes.[2] The definition and significance of arbitrage, for instance, are introduced in the first chapter with a simple gambling example. The limitations of the "Efficient Market Hypothesis" are compared with the constraints of a quadratic Taylor series representation for $y = \cos(x)$.

---

[1]This is critical: After the 2008 crash, some government experts attended an NRC committee meeting to explore what they missed. A couple were surprised to discover that key equations from this area are not always applicable. When one wondered how to discover such information, I volunteered my better students.

[2]This appeal to general concepts avoids the common problem where you must know something about finance before you can study (or teach a course) in this area.

Teaching always involves compromises between time and depth of coverage, such as in quarter length courses. My advice is to pace the course to ensure that the messages of Chapters 6, 7, and 8 are covered. Not doing so would be akin to reading an Agatha Christie mystery novel only to discover that the last chapters are missing.

Problems at the end of each chapter are roughly in the order of the chapter's presentation, which makes it easier for assigning homework. Coming up with other problems, by mimicking examples in the text, is easy. My students are responsible for all the problems in covered chapters, and my spot quizzes typically come from changing numbers in assigned problems.

Ideally, the course should emphasize how the power of mathematics significantly assists developing a sense, an intuition, about the market. When the material becomes mathematically more technical, there is a danger that students will focus on mathematical details at the expense of developing intuition about financial options. To counter this, "intuition breaks" are inserted in various locations. Many serve as reasonable homework problems.

Subject to time constraints, other features can be addressed. To suggest opportunities when teaching a course on dynamics, it is possible to appeal to Newton's law or blocks sliding down inclined boards. But for the uninitiated, much of economics is a mysterious world; instinct, experience, and intuition should be provided. OK. How?

Central to the material is the "buy low, sell high" phrase. A way to experience this cliché (time permitting) is with an experiment where a portion of the class, the *suppliers*, produce widgets: in fact, they buy them from an imaginary "Sue." A different portion of the class are the *buyers or consumers*. The value each attaches to widgets is determined as follows: each consumer can purchase widgets from the suppliers and sell them to me.

Reflecting different levels of manufacturing expertise, Sue offers different students different prices (as specified on slips of paper that the students draw from a bowl); they have no idea what prices the other suppliers have. Similarly, each buyer draws a slip from a bowl which states what I will pay for a widget; different buyers have different prices to reflect the different values consumers place on widgets.

The market opens as soon as some student offers to buy, or sell, a widget at a price they specify: if someone agrees, a sale is made, and the two are out of the market. Each person marks on the slip the sale price to determine personal profits. (The instructor's role is to explain the process, ensure that the auction starts, and maintain order—whatever happens with the bidding happens.) Everything continues until no more sales can be made. For instance, Sue (i.e., the slips of paper drawn by the students who are suppliers) may offer each of six students a price from 2, 4, 5, 7, 9, and 10; similarly, my offers may come from 3, 5, 6, 6, 8, and 9. (Efficient group sizes range from 25 to 30 each; they need not be the same size.) No real money is involved, but students, even observers, quickly become captivated.

This game is carried out several times (there is a learning process; in later rounds, students become more strategic and sophisticated) where each round uses different prices. (So, if a student with a raw deal ends up in a subsequent round, she or he may draw a better choice.) The message is when the supply and demand curves from each experiment are plotted on the board, the intersection is close to the price obtained through the games: not only does this exercise provide validity for the supply versus demand story (for particular settings), but the bidding process demonstrates how the "wisdom of the crowd" influences the search for an equilibrium price.

This course is enjoyable to teach! Of help for readers are my YouTube lectures found under "Math 176, Mathematics of Finance," which cover most of the material in this book. As for required background, students who have finished the calculus sequence (several variables) and an introductory course in probability and statistics have been successful.

Finally, my thanks to Dan Jessie for corrections and suggested changes in the notes after he taught this course several times. Santiago Guisasola modified portions of these notes to teach gifted high school students in a summer camp. My thanks to Anneli Duffin and Katri Sieberg for their assistance during the development of this material. Thanks to the five reviewers for their useful comments. And, in particular, my thanks to the many students for their feedback!

# Introduction

A delight surrounding the *Mathematics of Finance* is that while much is known, so much is unknown. Consequently, with the current state of understanding, it is wise to treat the models to be discussed as first-cut attempts of providing structure, sense, and some science to that vast amount of data and uncertainty that characterizes financial interactions. With time, these expressions will be, and are being, improved. The message is to learn to be skeptical: *don't fully accept anything.*

As true for any area characterized by a lack of a mature understanding, there is a sense of excitement generated by the inherent dangers and the hidden opportunities. The dangers reflect the reality that serious errors can be, and are being, made. In particular, anticipate errors whenever existing models are applied in inappropriate settings. (This happens!!) Remember, in finance, an error can translate into a significant loss of money.

Opportunities exist because so much of the unknown is waiting to be discovered. To exploit these opportunities, limitations of existing models must first be understood. Doing so requires knowing when and why a particular model can provide wrong answers and where caution is advised—this requires understanding the mathematics behind the finance. Can, for instance, models be corrected, extended, or changed to more accurately handle emerging challenges?

Of particular importance is the need to develop *intuition* about what might happen and why. After all, should a financial opportunity arise, it most surely will not wait for you to run home, pull out a book, find the appropriate equation, and then compute an answer. Intuition, developed by questioning why certain conclusions are true, by understanding when and why certain approaches are applicable, can help make a rapid response.

To handle the challenges of this area, develop the habit of *critical thinking.* Rather than accepting given assumptions—following the traditional approach expected in most courses where a student fully accepts what is said and then concentrates on details—carefully and critically examine all assumptions. Take the attitude of:

- Can these assumptions be believed?
- Do they make sense?

- When and where are they valid? When and where can they be *wrong*?
- What impact do these assumptions have on the conclusions?
- Would different assumptions lead to conclusions more consistent with what is observed?
- What happens should assumptions be modified?

Remember, a model is a mathematical attempt to understand or, at least, approximate reality. When differences occur between theoretical predictions and observations, when data keeps pushing forth contrary messages, be willing to suspect that the model, not reality, is at fault.[1] Thus, there is a need to continually compare models with actual events. Develop a habit of reading the financial and business news.

The reader might be wondering: "How can I construct or improve mathematical models?" Simply take advantage of whatever mathematical and economic tools you understand and can use. The more tools and the more mathematics and economics a person knows, the better equipped she or he is to make advances. Again, "critical thinking" is required. To employ mathematics and economics in new, novel ways, there is a need to go beyond knowing what is found in standard textbooks.

- Develop intuition about what kinds of mathematics can and should be used in different settings.
- Understand how and when particular mathematical results are overly restrictive for our needs.
- Become sensitive to the kinds of general results that need to be created to accomplish *your* goals.

An important element implicit through this book is the need to develop skills in critical thinking.

- Learn how to separate details from concepts. Here is an easy test: details are technical specifics of a field, while concepts tend to be more general; they apply to a variety of simple, everyday examples.

  An excellent test of whether concepts are being mastered is the ability to explain a newly learned concept to a sibling, a friend, or someone who is not familiar with this topic. If this can be done successfully, the concepts probably are being grasped, and the true meaning of the models are better understood. This approach offers a chance to understand where and why assumptions are reasonable tools to advance our understanding and where and why they may be lacking. If such stories cannot be told, there is the danger of becoming a slave to the technical details and models proposed by others.

---

[1] As this comment is obvious, it is surprising how even experts can forget it. It is not overly difficult to find examples where experts brand people as being "irrational" because their actions fail to match theoretical predictions. The people are not at fault; it is the model.

- Learn how to evaluate assumptions and models. The most important question is **WGAD**?[2] Assume a pragmatic stance of critically evaluating an approach, a set of assumptions, with others.
- To understand how and what kind of mathematics may be useful, acquire an intuitive sense about what each type of mathematics offers. To develop intuition, tell stories about the different topics to mathematically challenged friends. If the concepts can be explained, there is reason to believe you understand the mathematics.

Throughout this book, an attempt is made to illustrate these points. Several of the important concepts are introduced with stories involving commonly understood behavior. Rather than assuming the reader recalls needed mathematical results, many are reviewed with, again, stories. Some exercises are designed to help you develop intuition about what should happen.

Only so much can be covered in any course. And so, this book provides a mathematical introduction into a portion of that broad and fascinating area of finance. There is so much more to explore. In doing so, let me emphasize, again, the importance of developing intuition and to start paying attention to the business pages.

And now, enjoy the material!

---

[2]Who gives **a darn**!

# Contents

# Chapter 1
# Preliminaries via Gambles

## 1.1 A Football Game

Before tackling the complexities of the financial market and encountering unfamiliar words such as "options," "hedging," "arbitrage," "Puts," and "Calls," consider a simpler issue that, in fact, captures much of what will be discussed. Suppose next Sunday there will be a football game between the Vikings from Minneapolis and the Packers from Green Bay.

It is not clear which team has an advantage, but Bob, an avid Viking fan, is so confident about his favored team that he offers 25 to 1 odds.[1] This means that:

- If the Vikings win, Bob collects all of the money bet against him.
- If the Packers win, then for each dollar bet against him, Bob will pay $25.

No money changes hand until the game is over.

Sue supports Green Bay but with more restraint as manifested by her offered 6 to 5 odds. Again,

- If the Packers win, Sue wins all of money bet against her.
- If the Vikings win, Sue pays $6/5 = $1.20 for each dollar bet against her.

To make this personal, suppose you, the reader, have only $100 where every single penny is urgently needed to buy course books. Should you bet this money? If so, with whom?

Beyond the risk inherent in gambling with desperately needed money, what makes this problem challenging is that many readers will know nothing, nor even care, about football! Others may know there is such a game, but know nothing about

---

[1]Outrageous! But this is the actual choice offered by a student during a course!

© Springer Nature Switzerland AG 2019

D. G. Saari, *Mathematics of Finance*, Undergraduate Texts in Mathematics,
https://doi.org/10.1007/978-3-030-25443-8_1

the abilities of these two teams. And then those with some knowledge need not be experts. Consequently, no matter what bet might be made—with Bob or with Sue—it most surely is accompanied with uncertainty and risk. Similar to how some people buy stocks on the market, bets may even be based on emotion and instinct rather than careful analysis.

### 1.1.1  Removing Uncertainty

On the other hand, maybe, just maybe, it might be possible to remove the uncertainty from such wagers. Is there a way to create a "sure thing" by betting in a sufficiently clever style to ensure a profit *no matter which team wins the game?*

Sounds impossible. But should such a strategy exist, the scheme clearly must involve betting on both teams. Namely,

- Bet some money on the Packers with Bob.
- Bet the rest of the money on the Vikings with Sue.

The problem is to determine the appropriate split—the correct amounts to wager on each of the two possible outcomes. Later, after introducing financial terms, this strategy of betting on both sides of an issue as a calculated way to reduce or eliminate risk will be called *hedging*. Among the various definitions, the one used here is:

**Definition 1** Hedging *is taking a contrary position (making an investment) to offset and balance the risk associated by assuming a position in a market.*

Sounds complicated, but it is not. Hedging merely is a strategy to minimize an investor's exposure to market changes. Hedging, for instance, permits supporting new ventures. Suppose Tatjana is evaluating whether to invest in a new technology, which could thrive or take a dive. She may be more inclined to make an investment if she can minimize her losses should disaster hit, so Tatjana would seek ways to hedge her bet.

Hedging provides a level of "insurance" to reduce risk. This "insurance" is achieved by betting on both sides of what the future may reveal. In fact, not only is hedging commonly used, but the reader probably has been involved: Car insurance, for instance, pays off—it rewards you—only if you have an accident. We drive carefully to avoid accidents, but insurance covers consequences should a mishap occur. As such, buying car insurance, which is betting on both sides of having an accident or not having an accident, covers whatever may happen. For this reason, hedging plays a central role in the material starting in the next chapter. But while hedging can provide protection, it can be, and has been, abused.

### *1.1.2   Computations*

To determine how much of the $100 our fearless gambler should bet with Bob and with Sue, let the gambler bet $x on the Packers with Bob and the rest, $(100 − x)$, on the Vikings with Sue. The problem is to determine

- the value of $x$ and
- the ensured amount of profit.

  Only two events can occur: Either the Vikings or the Packer will win.

- *If the Packers win.*
    Bob loses so he must pay our gambler $25x$.
    But Sue wins, so our gambler must pay her $(100−x)$. (Remember, no money is exchanged until the game is over.)
    Therefore, our gambler's earnings should the Packers win are

$$25x - (100 - x) = 26x - 100. \tag{1.1}$$

Money is earned if and only if $26x > 100$ or if and only if

$$x > 100/26 = 3.85. \tag{1.2}$$

- *If the Vikings win.*
    Sue loses, so she pays our gambler $\$\frac{6}{5}[100 - x] = 120 - \frac{6}{5}x$.
    Bob wins, so our gambler pays him $x$.
    Therefore, the total earnings should the Vikings win are

$$[-x] + [120 - \frac{6}{5}x] = 120 - \frac{11}{5}x, \tag{1.3}$$

which has a positive outcome if and only if

$$\frac{5}{11}(120) = 54.55 > x. \tag{1.4}$$

By combining both inequalities, it follows that any $x choice satisfying

$$3.85 < x < 54.55 \tag{1.5}$$

simultaneously satisfies Equations 1.2 and 1.4. Thus, any $x value offered by Equation 1.5 to hedge on this bet—betting an appropriate amount on each team—*results in a profit no matter which team wins the game.*

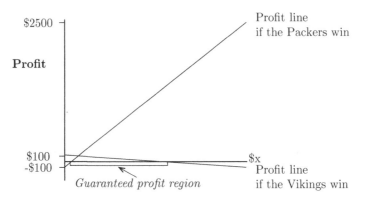

**Fig. 1.1** Profits from betting

### 1.1.3   Profit Curves

The profits accruing to the various $x$ choices are displayed in Figure 1.1. The interval described in Equation 1.5 are those $x$ values where both profit lines provide positive values, so they are in the region where both lines are above the x-axis.

Figure 1.1 identifies various options. Should our gambler believe that Bob is intemperate, or wrong, our gambler might bet on the high side of this $3.85 < x < 54.55$ region, which is where the profit line for the Packers is much higher; e.g., bet $50 on each team. Remember, each line's $y$-value specifies the profit earned should the indicated team win.

This x = $50 choice is a limited gamble because profit is guaranteed no matter what happens. But the difference can make a huge difference: Should the Packers pull it off, then, according to Equation 1.1 and Figure 1.1, our gambler would waltz away with $26 \times 50 - 100 = \$1,200$ in a risk free wager! Should the Packers lose, the measly $10 winnings may not even cover the cab fare home. The point is that the added information from curves of this type provide guidance.

### 1.1.4   Fixed Income

Rather than "betting" to enhance the excitement of the football game, suppose these wagers are part of a vocation. Here our gambler has no interest in random earnings, she wants to wager in a manner to ensure a *fixed* return no matter what happens during the game.

To achieve this goal, notice from Figure 1.1 that the highest assured profit is where the two profit lines cross. To find this profit value, set the two "earnings" (from Equations 1.1, 1.3) equal to obtain

$$26x - 100 = 120 - \frac{11}{5}x,$$

or

$$28\frac{1}{5}x = \frac{141}{5}x = 220.$$

The height of this point where the two profit lines cross provides the largest guaranteed winnings.

The conclusion of this algebra is that by betting $\$x = \$\frac{220\times 5}{141} = \$7.80$ on Green Bay and $\$100 - x = \$92.20$ on the Vikings, our gambler is ensured the earnings of

$$26 \times 7.80 - 100 = \$102.84$$

*independent of which team wins the game.* Remember, this profit is in addition to the starting $100 of our gambler; it provides more than a 100% return on our gambler's investment of $100.

An unexpected conclusion is that the amount bet on Green Bay, where the odds are tremendous, is much less than the amount to be bet on the Vikings. As it should be expected, Figure 1.1 provides an explanation. The greater odds that Bob places on Green Bay generates a profit line with a much higher slope. To ensure a guaranteed amount of winnings, where the lines cross, the $x$ value must be smaller.

### 1.1.5   Arbitrage

This sizable return is guaranteed without our gambler spending, risking, or putting forth even a dime! This is because, in the example, no money is exchanged until after the game. Even a gambler with an empty wallet can handle the exchange by first collecting the winnings to pay off the debt, and then return home with $102.84.

This sense of betting on all sides to ensure a fixed return is called *arbitrage*. As with the football example, arbitrage is where different prices for the same commodity on different markets make it possible to instantaneously "buy low, sell high." The "instantaneous" modifier implying there is no time gap is crucial because it eliminates exposure to market changes and risk.

**Definition 2** Arbitrage *is a strategy of ensuring a profit by simultaneously buying and selling the same asset in different markets in order to take advantage of the asset's different prices.*

Here is an obvious question: Why doesn't everyone adopt this strategy? A first answer is that such circumstances do not always exist. Then, when arbitrage is possible, many would like to take advantage of the setting—if they could. Doing so requires *knowing* that the opportunity exists and how to exploit it.

Realistically, not everyone recognizes when such opportunities arise: As an illustration, probably most readers failed to recognize the financial opening provided by this football example. Stated in blunt terms, to the observant and clever go the profits: Market opportunities exist only for those who recognize how and where they can be found. And so, throughout this book, the reader is encouraged to sharpen personal instincts and intuition by checking whether and where arbitrage opportunities exist.

Attached to arbitrage is the "money pump." To describe this with the Bob and Sue arbitrage opportunity, those who understand how to take advantage of the setting might be tempted to "pump" money out of it. With more and more people exploiting this opening, Bob and/or Sue will quickly learn either to be more cautious or to suffer the consequences. In particular, expect their offered odds to quickly adjust until the inefficient arbitrage opportunities either disappear or become minimal.

"Inefficiency" is key; savvy investors may seek markets that are known for inefficiencies when seeking arbitrage advantages. It may involve a time delay in transactions or lack of information about the market. Admittedly, this activity has an unsavory, negative overtone, suggesting the need for a long hot shower. But, as with Bob and Sue, its potentially positive effect is to force markets to adjust to more realistic, efficient levels. In mathematical terms, the adjustment mechanism of arbitrage is what forces markets to adjust, and then adhere, to appropriate, more stable settings.

It is worth restating this last comment. Of importance to the mathematical theory is that this arbitrage functioning is what permits us to assume that certain functions and outcomes in the financial world are "continuous" or "differentiable." As a comparison, in the physical sciences, nature's adjustments provide the rationale for assuming that functions are smooth (that is, differentiable as many times as needed); in finance, it is the opportunistic shoving and pushing of arbitrage that supplies the justification.

With that said, expect arbitrage to be central to much of what follows: It is what permits accepting invariants, such as the valuable "Put–Call Parity Equation" (introduced in the next chapter) as well as smoothness assertions for mathematical expressions. The possibility of arbitrage is what introduces adjustments to handle appropriate kinds of equilibrium.

### 1.1.6   Hedging Without Arbitrage

The football example unites hedging with arbitrage; typically they are separate. To illustrate, our gambler may wish to take advantage of Bob's tempting odds. But to have enough money to return home, she cannot afford to lose more than $10. What should she do?

She could bet $10 with Bob, and only with Bob. If the Packers win, she goes home with the sizable bonus of $250. If the Packers lose, she loses the full $10.

There is another option; she could hedge by also betting with Sue. Here she would bet $y on the Packers with Bob and $(100 − y)$ on the Vikings with Sue. What differs from the previous analysis is that her bet on the Vikings serves as insurance to ensure that she loses no more than $10. So, if the Vikings win, she pays Bob $y and earns $\frac{6}{5}(100 − y)$ from Sue (see Equation 1.3) for the return of $-y + [120 − \frac{6}{5}y]$.

Remember, she is willing to take a ten dollar loss to enhance the amount that she can bet on Green Bay. Thus, the appropriate equation for $y is

$$-y + [120 − \frac{6}{5}y] = -10,$$

which is what she can lose, or

$$130 = \frac{11}{5}y; \quad y = \frac{650}{11} = 59.09.$$

This means she should bet $y = 59.09$ on the Packers with Bob, and the rest with Sue; doing so ensures losing no more than $10 no matter what happens. Rather than betting only $10 with Bob, she can significantly increase her bet to $59.09 and remain within her budget.

If the Packers win, the profit comes from Equation 1.1 by substituting this $y = 59.09$ value. With this victory, she dances all the way home with

$$26\frac{650}{11} − 100 = 1,436.36.$$

A powerful return of over a thousand dollars is achieved with the risk of losing only $10! OK, it is highly unlikely to encounter such an extreme setting; presumably Bob has learned his lesson and no longer offers such odds. But this example demonstrates how hedging provides insurance by minimizing risk and losses while offering opportunities. Think in terms of how it allows an investor to invest more money in a new venture.

### 1.1.7  A Cautionary Word of Interpretation

A word of caution: this hedging and arbitrage analysis is described in terms of profits; it describes a setting among trusting individuals. The expectation is that if you lose, then, as fully accepted, you will pay. There is no need to rely on "collectors" with hairy knuckles deploying baseball bats.

Settings that do not enjoy personal connections cannot survive on such trust. One would not expect to see a flood of anonymous betters grasping losing tickets while rushing with their money to pay what they owe. Where trust is questionable, bets are collected *in advance*. If you win, you will get it back. With Sue's 6:5, or 1.20:1 odds, what may be expressed is that you bet $1 bet and win, then you will receive

your $1 back plus your profit of $1.20 or $2.20.

With actual examples, be sure to determine whether the description is in terms of profits or in terms of profits plus the money put up in advance. The results differ significantly should attention not be paid to this distinction.

Experts learn quickly, so rarely do settings provide an arbitrage opportunity. Yet it is reasonable to expect non-experts will wager over emotional events, such as the (fictional) upcoming election between the Green and Orange parties. To ensure payment at the end, when Tyler places a $1 bet with Mikko that the Greens will win, he will receive $1.50 if the Greens succeed (so, this is Tyler's $1 guarantee plus the $0.50 profit) and if Tyler places a $1 bet with Jane on the Orange party's success, he will receive $1.80 if the Orange group wins (but only $0.80 profit). If Tyler used the above analysis without paying attention to profits, it will appear there are strong arbitrage opportunities. In reality, there may be hedging options, but there are no arbitrage possibilities.

To see why, consider Jane, a Green party supporter: After removing what it cost to buy the bet and concentrating only on profits, her offered odds are $\frac{1}{2}$:1. Similarly, Mikko supports the Orange party and he really is offering $\frac{4}{5}$:1 odds. What follows is the analysis where Tyler has $1000.

If, with Jane, Tyler bets $x on the Orange party and bets the rest, $1000 − x, with Mikko on the Greens, the analysis is as follows:

- If the Greens win, Tyler loses the $x bet with Jane, but wins $\frac{4}{5}[1000 − x]$ dollars from Mikko for winnings of

$$-x + \frac{4}{5}(1000 - x) = -\frac{9}{5}x + 800. \tag{1.6}$$

To ensure a profit when the Greens win, it must be that

$$\frac{4000}{9} \approx 444.44 > x. \tag{1.7}$$

- If the Orange party wins, then $(1000-x) is lost to Mikko, but Jane must pay $\frac{1}{2}x$ leading to the earnings of

$$-(1000 - x) + \frac{1}{2}x = \frac{3}{2}x - 1000,$$

where, to have a profit should the Orange party win, it must be that

$$x > \frac{2000}{3} \approx 666.67. \tag{1.8}$$

- So, by betting any $x on the Orange party where

$$444.44 < x < 666.67$$

and the rest of the $1000 on the Greens leads to a *guaranteed loss* no matter which team wins!! Ouch!

An immediate issue is to understand when arbitrage opportunities arise, and when risk and loss prevail. The analysis starts next.

## 1.2  Expected Value and Variance

It is reasonable to wonder why a discussion about the mathematics of finance opens with a betting description. To provide an answer, consider whether you should, or should not, buy Apple stock today. (Remember, buying a stock means you are buying a unit of ownership of the company.) Doing so is a gamble; it is a bet. Consequently, certain principles that are easier to understand with gambling stories, such as the football or election narrative, transfer to appreciate subtle issues of finance.

To decide what needs to be known whether to purchase that stock, appeal to instinct. Knowing there is a 70% chance Apple stock will increase is valued knowledge. To appreciate how to use this information, a review of the appropriate mathematics of "likelihood" is needed. What follows is a swift reminder of basic concepts from probability and statistics; it is based on an expectation that the reader has at least an elementary understanding of these topics, so only an outline of terms and notions is necessary. To make the discussion inviting, some unusual examples are provided.

### 1.2.1  Probability and PDF

Suppose there are $n \geq 2$ mutually exclusive events where $p_j$ is the probability that the $j^{th}$ event will occur where, of course, $p_j \geq 0$. The fact that one event must occur leads to the standard constraint

$$\sum_{j=1}^{n} p_j = 1. \tag{1.9}$$

**Examples**

**Spinning Coins.** When a coin is spun on edge, the only possible events are Heads or Tails.[2] The probability of getting H, or T, denoted by $p(H)$ and $p(T)$, when flipping a coin, is essentially $\frac{1}{2}$ for each event. Surprisingly, these values differ significantly for a spinning coin.

The reason is that should one side of the coin be slightly heavier, it will tilt the axis of rotation making that face more likely to end up on the bottom. With an older American penny, the heavier Head makes it more likely for Tails to come up. After considerable experimentation from students in several classes, we learned that

$$p(H) = 0.3, \quad p(T) = 0.7. \tag{1.10}$$

These results are based on using pennies from prior to 2000;[3] I don't know what happens with more "modern" coins. Try it; spin a penny, or, because pennies are rarely used, maybe a quarter, to determine the approximate probabilities for the specified coin.

**Dice.** The likelihood of each face of a regular six-sided die occurring is $\frac{1}{6}$. Each $p_i$ is greater than zero and, trivially, the sum of these six values of $\frac{1}{6}$ equals unity. So, the likelihood of rolling a six or a three is the same; both equal $\frac{1}{6}$.

With two dice and the 36 possible outcomes for the combination of the red and blue die, the answer changes depending on how the question is phrased. If the question is to find the likelihood that the red die is one and the blue one is six, the answer is $\frac{1}{36}$—there are 36 possibilities and only one choice satisfies the specified condition. If the question is to find the likelihood that a seven is rolled, the answer changes. Carry out the details by making a table of all 36 outcomes. (Six of these outcomes sum to seven, so the likelihood is $\frac{6}{36} = \frac{1}{6}$.)

**Gender of Children.** Assume it is equally likely for a newly born child to be a boy or a girl, which means that $p(B) = p(G) = \frac{1}{2}$. With two children in a family, the *sample space*, or space of possible events, is

$$\{BB, \ BG, \ GG, \ GB\}, \tag{1.11}$$

---

[2]An unlikely outcome has the coin stopping on edge: Once, while with faculty from Boston University after giving a colloquium talk, I spun a penny to illustrate the unexpected H and T values described in this example. To everyone's surprise, the penny stopped—on edge! Now that this "once in a lifetime" event is behind us, treat "landing on edge" as having a zero probability. That is, if E represents "edge," then p(E) = 0. As this description illustrates, the probability of an event being zero need not mean it is impossible. It only means it is smaller than any imagined positive number, no matter how small.

[3]After a class of mine discovered these values, I learned that some of the students used this fact to increase their discretionary funds.

where the ordering specifies the birth order; e.g., $BG$ means that a boy was born first and a girl second. The importance of this ordering will become apparent.

A common expectation is that each of these four Equation 1.11 events is equally likely. *Is this true?* Data confirms that it is reasonably accurate in the aggregate. But as experience already with course grades proves, results about the aggregate can differ from assertions about an individual. So, what are the probabilities for a particular family? What underlying assumptions would ensure this condition? Whatever are the appropriate requirements to establish that each event is equally likely, it follows that

$$P(BB) = P(BG) = P(GB) = P(GG) = \frac{1}{4}.$$

Suppose the goal is to compute the likelihood a family has one boy and one girl. This event, $BG \cup GB$, which means that either a boy or a girl is the first born, and the second child is of the opposite gender. As the events are disjoint, it follows that

$$P(BG \cup GB) = P(BG) + P(GB) = \frac{1}{4} + \frac{1}{4} = \frac{1}{2}.$$

The care taken in listing the order of births makes this a simple computation.

An important aspect of finance is to understand how to use *information* to your advantage; information may be useful even it appears to be extraneous. To illustrate with this two child example, suppose Jacqueline is invited to have dinner with a friend and her family. All Jacqueline knows is that her friend has two children. Consider two possibilities:

1. When Jacqueline rings the doorbell, a delightful, polite, little girl answers and says, "Hello. My name is Anni. Please come in."
2. When Jacqueline rings the doorbell, Amy, a snotty little girl, answers and says, "Go away, I don't want anybody to come here! And I always get my way because I am the youngest."

For each case, compute the likelihood the other child is a boy. Does the answer change? Surprisingly, it does.

All Jacqueline knows with the first scenario is that one of the two children is a girl, which means that her friend does not have two boys. Consequently, rather than Equation 1.11 sample space, the actual sample space is

$$\{\overline{BB},\ BG,\ GG,\ GB\} = \{BG,\ GG,\ GB\},$$

where $\overline{BB}$ means that the event is eliminated. Each of the three remaining events is equally likely, so each now has the probability of one-third. The only way the other child could also be a girl is with $GG$. The two remaining possibilities have a boy as the other child, so the answer is two-thirds.

The second scenario's relevant information has nothing to do with the girl's sour attitude; it is that the *youngest* child is a *girl*. Therefore, neither $GB$ nor $BB$ (where a boy is the youngest) are admissible events; the actual sample space is

$$\{\overline{BB},\ BG,\ GG,\ \overline{GB}\} = \{BG,\ GG\}.$$

Both events in this reduced space are equally likely: $P(BG) = P(GG) = \frac{1}{2}$, so the likelihood the other child, the older one, is a boy is 0.5.

Here is a question; suppose the girl announced that she was the oldest? Does the answer change?

*Comment:* The above involves conditional probability. The reader is advised to review conditional probability and then relate it to the above examples.

**Examples Involving Integrals.**  Discrete probability models are not sufficient for many financial settings. To handle the more extensive distributions, turn to the power of integral calculus.

Recall that the value of an integral, $\int_a^b f(x)\, dx$, can be interpreted as the area under the curve $y = f(x)$ over the interval $a \le x \le b$. (This is an interpretation, *not* a definition.) To illustrate with the function $y = f(x)$ depicted in Figure 1.2a, this $\int_a^b f(x)\, dx$ value represents the area between the curve and the $x$-axis.

A way to compute this Figure 1.2a area is to approximate the integral's value. As the reader most surely recalls from an introductory course in calculus, to do so, partition the interval $[a, b]$ into $n$ subintervals as indicated in Figure 1.2b.

Select these subintervals to have an equal width of

$$\Delta x = \frac{b - a}{n}.$$

On each subinterval, form a rectangle to approximate the area under the curve in this region. The height of the rectangle is given by the $f(x_j)$ value for some point

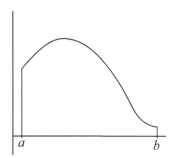

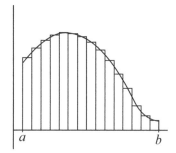

**a.** The continuous distribution           **b.** A discrete approximation

**Fig. 1.2**  Continuous distribution

$x_j$ in the subinterval, which leads to the rectangle area (of height times width) of $f(x_j) \times \Delta x$. By adding the areas of all the regions (as represented in Figure 1.2b), an approximate value is given by

$$\text{Area} \approx \sum_{j=1}^{n} f(x_j)\, \Delta x.$$

The sum of the depicted rectangular areas offers a reasonable approximation for the area, but it is possible to do much better. For a sharper answer, use more intervals and rectangles; i.e., choose a larger $n$ value.

The *definition* of the integral is the limiting value obtained if this process could be continued indefinitely. Namely, the definition is

$$\lim_{n\to\infty} \sum_{j=1}^{n} f(x_j)\, \Delta x \to \int_{a}^{b} f(x)\, dx. \tag{1.12}$$

(assuming that the limit exists).

This approach shows how to replace a complicated, continuous probability density function with a finite discrete approximation! Just as in Figure 1.2b, replace the continuous model with a discrete approximation defined over the small $\Delta p$ intervals: The probability is approximated by summing the various $f(p)\Delta p$ rectangle areas, where $f(p) \geq 0$. With this integral formulation, Equation 1.9 constraint assumes the representation

$$\int_{-\infty}^{\infty} f(p)\, dp = 1. \tag{1.13}$$

Function $f(p)$ (or, if you wish, the value $f(p)\Delta p$ leading to $f(p)\, dp$) is called the *probability density function* or *PDF*.[4]

To illustrate, the probability of randomly selecting a number that lies in a specific subinterval of $[0, 3]$, where each value is equally likely, has a constant PDF. Thus,

$$f(p) = \begin{cases} 0 & \text{if } -\infty < p < 0, \\ C & \text{for } 0 \leq p \leq 3 \\ 0, & \text{for } 3 < p < \infty. \end{cases} \tag{1.14}$$

The value of $C$ is determined from Equation 1.13; this means that

---

[4]The original notation was "pdf," which now is more commonly associated with something else in common use.

$$\int_{-\infty}^{\infty} f(p)dp = \int_{0}^{3} C \, dp = 1, \quad \text{or } C[3 - 0] = 1; \quad \text{thus } C = \frac{1}{3}.$$

Armed with the PDF, the probability of selecting a number in a specified subinterval can be computed. The probability that the selected number is between 0.45 and 0.78, for instance, is

$$P(0.45 < x < 0.78) = \int_{0.45}^{0.78} \frac{1}{3} \, dp = \frac{1}{3}[0.78 - 0.45] = 0.11.$$

### 1.2.2   Random Variable

A random variable is a function, which differs from the usual $y = x^2$ variety in that its inputs—its domain—consist of the random events that can occur. In spinning a penny, the two events are Heads and Tails.

The random variable's range specifies what happens with each particular random event. For instance, flipping a penny to determine whether or not to watch the movie might have the consequences

$$X(x) = \begin{cases} \text{Watch the movie} & \text{if } x = H; \\ \text{Don't watch the movie} & \text{if } x = T. \end{cases}$$

A common example is where the outcome is determined by the random event, such as whether the weather is, or is not, nice. The random variable $X$ could be

- $X$(bad weather) = {go to class}
- $X$(good weather) = {skip class; go to the beach}

Standard choices of a random variable $X$ identify a number with each event. One example has $X(j)$ representing the "winnings" that occur when the $j^{th}$ event occurs. Another choice with the roll of a die is where the random variable just identifies the number on the die's top face.

### 1.2.3   Expected Value

Loosely speaking, when the outcomes of a random variable $X$ are numbers, its *expected value*, denoted by $E(X)$, is what can be "expected"—it is a form of an average. To illustrate, consider the expected winnings from spinning a coin where $p(H) = 0.4$ and $p(T) = 0.6$ and the winnings are $10 if Heads occurs, but a loss of $10 with Tails. Because $H$ can be expected to occur 0.4 of the time, the expected winnings should be

$$0.4 \times \$10 = \$4,$$

while the expected losses are

$$0.6 \times -\$10 = -\$6,$$

for a total of

$$\$4 - \$6 = -\$2.$$

More generally and precisely,

**Definition 3** *Let X be a random variable with $f(p)$ as the PDF. The expected value of X is*

$$E(X) = \int_{-\infty}^{\infty} X(p) f(p) dp. \tag{1.15}$$

*If the probability distribution is discrete, then*

$$E(X) = \sum_{j=1}^{n} X(j) p_j.$$

**Example** In Equation 1.14 example of randomly selecting a number from the unit interval $[0, 3]$, the random variable $X(t) = t$ specifies the selected number. The *expected value* of $X$ is

$$\int_{-\infty}^{\infty} X(t) f(t) dt = \int_{0}^{3} X(t) \frac{1}{3} dt = \int_{0}^{3} \frac{t}{3} dt = \frac{t^2}{6} |_{0}^{3} = 1.5. \tag{1.16}$$

This makes sense; 1.5 is the midpoint.

## 1.2.4   Variance

It is fair to identify the expected value with the "center of mass." Of particular interest, the expected value is a natural reference point from which to measure something. Stating that a test score is 80, for instance, fails to indicate whether it is good or bad. Learning that it is 10 points above the expected value (the average) implies that it is good. But, how good?

Answering the "how good" concern requires a unit of measure. As an introduction, suppose the only store that carries a needed course book is seven away from here. "Seven?" The statement that the store is seven away means nothing. Seven "what?" Seven meters? Seven blocks? Seven kilometers? Seven miles? Seven states? The choice of units distinguishes between walking or hitching a ride. A step toward creating a unit of measure for data is the definition of "variance."

The following description borrows heavily from the usual $\sqrt{x^2 + y^2}$ distance definition. In fact, this expression is used to find the "distance of data" from the expected value.

**Definition 4** *For a random variable X with expected value $\mu = E(X)$, the* variance *is*

$$\text{Var}(X) = E([X - E(X)]^2) = E([X - \mu]^2).  \tag{1.17}$$

*For a discrete probability, Equation 1.17 becomes*

$$\text{Var}(X) = \sum_{j=1}^{n} [X(j) - \mu]^2 p_j.  \tag{1.18}$$

*For a continuous probability distribution with PDF f(t), Equation 1.17 becomes*

$$\text{Var}(X) = \int_{-\infty}^{\infty} (X(p) - \mu)^2 f(p) \, dp.  \tag{1.19}$$

Equation 1.18 resembles the $\sum_{i=1}^{n}(x_i - a_i)^2$ equation for finding the square of the distance of a point $\mathbf{x} = (x_1, \cdots, x_n)$ from a specified location $\mathbf{a} = (a_1, \cdots, a_n)$ in a $n$-dimensional space. (Readers familiar with concepts from astronomy and general physics will recognize this expression as the "polar moment of inertia" relative to the center of mass.) The actual distance between $\mathbf{x}$ and $\mathbf{a}$ is the square root, or $\sqrt{\sum_{i=1}^{n}(x_i - a_i)^2}$. Similarly the unit of data measurement of "distance from the expected value" is the square root of the variance; it is the *standard deviation*.

**Definition 5** *The* standard deviation *of a random variable X is given by*

$$\sigma = \sqrt{\text{Var}(X)}.$$

For a quick way to compute the variance of a random variable, notice that because $(X - \mu)^2 = X^2 - 2\mu X + \mu^2$, it follows that

$$\text{Var}(X) = E([X - \mu]^2) = E(X^2) - 2\mu E(X) + \mu^2.  \tag{1.20}$$

This expression relies upon the facts:

- An integral of a sum is the sum of the integrals (i.e., $\int (f(x) + g(x))dx = \int f(x)dx + \int g(x)dx$, so $E(X^2 - 2\mu X + \mu^2) = E(X^2) - E(2\mu X) + E(\mu^2)$).
- A constant ($2\mu$ in the second integral) can be factored out of an integral.
- The expected value of a constant is the constant (here, $\mu^2$).

Because $\mu = E(X)$, Equation 1.20 becomes the more convenient

$$\text{Var}(X) = E(X^2) - (E(X))^2. \tag{1.21}$$

**Example** When randomly selecting a number from the interval $[0, 3]$ with PDF given by Equation 1.14 and $X(t) = t$, the $\mu = \frac{3}{2}$ value was computed. To determine the variance by using the definition, the integral

$$\text{Var}(X) = E([X - \frac{3}{2}]^2) = \int_{-\infty}^{\infty} [X(t) - \frac{3}{2}]^2 f(t)dt = \int_0^3 (t - \frac{3}{2})^2 dt$$

must be evaluated, which it can with a change of variables. However, Equation 1.21 permits the simpler computation of

$$E(X^2) = \int_0^3 \frac{1}{3} X(t)^2 dt = \frac{1}{3} \int_0^3 t^2 dt = \frac{t^3}{9} \Big|_0^3 = 3,$$

so

$$\text{Var}(X) = E(X^2) - (E(X))^2 = 3 - (\frac{3}{2})^2 = \frac{3}{4}.$$

Therefore, $\sigma = \frac{\sqrt{3}}{2}$.

**Example** When randomly selecting a number from the unit interval $[0, 1]$, a similar computation shows that $\mu = \frac{1}{2}$ and the PDF is $f(p) = 1$ for $0 \leq p \leq 1$, and zero elsewhere. Thus

$$E(X^2) = \int_{-\infty}^{\infty} X^2(t) f(t)dt = \int_0^1 X(t)^2 dt = \int_0^1 t^2 dt = \frac{t^3}{3} \Big|_0^1 = \frac{1}{3}.$$

This means that

$$\text{Var}(X) = E(X^2) - [E(X)]^2 = \frac{1}{3} - \left(\frac{1}{2}\right)^2 = \frac{1}{12},$$

leading to $\sigma = \sqrt{\frac{1}{12}} = \frac{1}{6}\sqrt{3}$.

Common examples using the mean (or expected value) and the standard deviation are IQ tests and SAT scores. One form of an IQ test adjusts the scores so that $\mu = 100$ and $\sigma = 15$. Therefore, a person with an IQ of 130 means that person's score on the test is 2 standard deviations above the mean. On some tests, a "genius" is treated as someone whose score is two or more standard deviations above the mean. Thus, rather than being someone special, a genius could be just a good "test taker."

Similarly, on the SAT tests, the mean is adjusted to be $\mu = 1000$, while $\sigma = 194$. A student with a 1200 SAT, then, scored slightly better than a standard deviation above the mean. For the ACT, the average composite score is 20.8 with a standard deviation of 4.8. So, a student with a composite 26 is slightly above one standard deviation.

### 1.2.5   Standard Form

Following up on the SAT and IQ tests, there is a standard representation of a random variable $X$. The first step captures whether the outcome is larger, or smaller, than the expected value, which suggests using

$$X - E(X).$$

The sign of $X - E(X)$ indicates if an outcome is above or below the expected value: Did you do better or poorer than the class average?

To impose a sense of distance, divide by the standard deviation $\sigma$ to obtain the *standard form* of $X$ of

$$Z = \frac{X - E(X)}{\sigma}. \qquad (1.22)$$

Thus the above SAT scores can be expressed as $1000 + 194Z$, while the ACTs are $20.8 + 4.8Z$.

A $Z = 2.5$ value for a random event not only is above the expected value (the sign is positive), but it is significantly above. On the other hand, a $Z = -0.5$ means it is below the average (the negative sign) by half a standard deviation. Using these values in more personal settings, a student receiving a 60 on the first exam where the average (expected value) is 85 did a horrible job if the standard deviation is 10, but not so bad if the standard deviation is 30.

Here is an application that might be dear to some readers. On the first exam, Ernesto's grade was right on the class average of 70, while Junying earned an 80; the standard deviation is $\sigma = 5$. On the second exam, Ernesto received a 90, while Junying's 80 was the class average with $\sigma = 10$. Are Junying and Ernesto tied in the class standing because each has an average test score of 80, or should Ernesto be ranked higher with his strong 90 performance on the second exam, or should Junying be ranked higher with her average Z score of 1 compared to Ernesto's 0.5? What is the more reasonable choice?

## 1.3   Fair Bets and Ensuring Profits

By using these notions, the reason profits can be ensured in certain wagers can be explained.

### *1.3.1   Fair Bet*

The notion of a *fair bet* means that the gamble, from either side, is equitable. It is so unbiased that the person offering the wager is willing to take either side because the expected winnings for either event prevailing are the same. Returning to the introductory Packers–Vikings game, if Bob offers what he views as a fair bet, then with these odds he would be willing to bet on either Green Bay or the Vikings.

**Definition 6**  *If random variable X represents winnings from a bet, the bet is "fair" if the expected value of the winnings is zero. That is, if*

$$E(X) = \sum_{j=1}^{n} p_j X(j) = 0. \tag{1.23}$$

All we know with the football example are the odds that Sue and Bob offer— these are the $X(j)$ payoff values. The likelihoods that each person assigns to the success of a team remain a mystery. But they can be computed if both believe they are offering a fair bet. This is because in addition to the fundamental Equation 1.9, which here is $p_1 + p_2 = 1$, Equation 1.23 introduces an extra equation with added information. With two equations and two unknowns, answers follow.

To illustrate, Bob's 25 to 1 odds define the random variable

$$X_B(V) = 1, \quad X_B(GB) = -25,$$

where Bob wins \$1 if the Vikings (V) win, but pays \$25 if Green Bay (GB) wins. Assuming Bob treats this as a fair bet, we can determine what Bob believes to be the probabilities of the Vikings (denoted by $p_B(V)$) and Green Bay ($p_B(GB)$) winning.

From the definition of a "Fair Bet" and Equation 1.9, it follows that

$$p_B(V)X_B(V) + p_B(GB)X_B(GB) = p_B(V)(1) + (1 - p_B(V))(-25) = 0, \tag{1.24}$$

or that $26p_B(V) = 25$. Therefore, Bob's *implicit probability,* or *subjective probability,* for the game is that the Vikings are highly likely to win because

$$p_B(V) = \frac{25}{26}, \quad p_B(GB) = \frac{1}{26}. \tag{1.25}$$

Similarly, Sue's implicit probability distribution for the game is

$$p_S(V) = \frac{5}{11}, \quad p_S(GB) = \frac{6}{11}. \tag{1.26}$$

A "Fair Bet" is introduced here because it is widely used and will be in what is discussed later. But for finding these personal subjective probabilities, all that is needed is a second equation. To illustrate this comment with Bob, a second equation could be, rather than a fair bet, he expects a $5 return for each $10 bet. Replacing Equation 1.24 would be

$$p_B(V)(10) + (1 - p_B(V))(-250) = 5, \qquad (1.27)$$

or $p_B(V) = \frac{51}{52}$ and $P_B(GB) = \frac{1}{52}$. Again, all that is needed is two equations with two unknowns; select whatever expression is realistic or convenient.

## 1.3.2   Making Money

If Bob and Sue offer what each views as a fair bet, why is it possible to make money off of them? To have an edge, to make guaranteed money, the gambler is *not* interested in a fair bet; the gambler seeks settings where Equation 1.23 replaces the zero with a positive value. But as this equation involves probabilities and the random variables $X(V)$, $X(GB)$, there are only two ways to convert the fair bet into a profitable setting—change the probabilities or change the winnings (the random variables).

Encountering different people who are willing to bet on the game with different odds is what can change the *gambler's* de facto probabilities. To see this, the probabilities *offered the gambler through the bets* are

$$p_B(GB) + p_S(V) = \frac{1}{26} + \frac{5}{11} = .4930 < 1, \qquad (1.28)$$

which grossly violates Equation 1.9. This inefficiency, this disregard for the laws of probability, these values that are not real probabilities introduce an arbitrage opportunity. And so the inequality of Equation 1.28 eliminates random effects; it ensures positive winnings no matter what happens. As we will see later, *eliminating random effects* is a prime motivation for the Black–Scholes Equations—an equation that dominates the area of finance.

Of course, if both Bob and Sue offered the same odds for each team, there would be no way to bet on both teams to ensure a profit. The reason is that Equation 1.28 would involve real probability values where the sum equals unity. More generally, arbitrage seeks settings where inefficiencies reside.

To underscore the point, return to the political example of the Orange and Green parties (Section 1.1.7). Jane offers odds of $\frac{1}{2}$ : 1, so if this is a fair bet, then $p_J(Green) + p_J(Orange) = 1$, while the fair bet expression (from her odds) is

$$p_J(Green) - \frac{1}{2}p_J(Orange) = 0.$$

Be solving these equations, it turns out that Jane is a pessimistic Green supporter because she believes there is only a

$$p_J(Green) = \frac{1}{3}$$

chance her party will be successful. Mikko is not so pessimistic about his party, the Oranges. Here we have $p_M(Green) + p_M(Orange) = 1$ with the fair bet value of

$$-\frac{4}{5}P_M(Green) + P_M(Orange) = 0.$$

Thus Mikko's sense of the Orange party winning is

$$p_M(Orange) = \frac{4}{9}.$$

Now turn to Tyler's plight. His wager on the Orange party with Jane has the implied probability $p_J(Orange) = \frac{2}{3}$, and on the Green party with Mikko has the implied probability of $p_M(Green) = \frac{5}{9}$. Instead of an Equation 1.28 inequality, Tyler encounters a reversed

$$p_J(Orange) + p_M(Green) = \frac{2}{3} + \frac{5}{9} = \frac{11}{9} > 1. \tag{1.29}$$

This violation of Equation 1.9 runs *against Tyler's interest,* which explains the lack of arbitrage opportunities (see Section 1.1.7).

To explain in a slightly different manner, suppose Jane and Mikko are agents for a betting group. If Equation 1.29 holds, then the betting group's probabilities are

$$p_J(Green) + p_M(Orange) = \frac{1}{3} + \frac{4}{9} = \frac{7}{9} < 1,$$

so this group's choices convert the expected outcome from a fair zero to a positive value.

### 1.3.3  Horse Racing

The message of the inequality in Equation 1.29 is that if a gambler confronts a setting where the sum of implied probabilities over all events equals or exceeds unity, there are no arbitrage opportunities. Expect such situations to arise when there are more than two possible events.

A natural example is the American roulette wheel, which offers a casino an advantage of slightly over 5% on gambles. A source of this advantage is the inclusion of a double zero green slot leading to a total of 38 slots. Rather than describing the computations in terms of gambling devices (which the reader might find to be an informative computation), a simpler example can describe what happens and why.

Suppose Zhihong, a hardworking, underpaid faculty member, needs to earn extra money on weekends, so he runs an off-track betting establishment. Today three horses, with the exotic names of A, B, C, are running: Zhihong must determine each horse's winning payoffs. Forget basing these payoffs on the abilities of the individual horses; base them on the amount of money that has been bet. Remember, this is a business: Zhihong could care less about which horse wins, but he needs to make 5 cents on each dollar bet *independent of what happens in the race*. Zhihong wants a 5% profit.

At the moment that betting is closed,[5] suppose $50 is bet on A, $30 on B, and $20 on C. Let the random variable $X(K)$ be the winnings per dollar bet if horse $K$ wins. By using this information, if A, B, or C wins, Zhihong's earnings are, respectively,

| Winning Horse | Zhihong's Earnings |
|:---:|:---:|
| A | $20 + 30 - 50X(A)$ |
| B | $20 + 50 - 30X(B)$ |
| C | $30 + 50 - 20X(C)$ |

$$(1.30)$$

Again, using nothing more than elementary algebra, the $X(K)$ values can be determined to ensure the desired fixed profit of 5 cents on each dollar wagered. With each $100 being bet, Zhihong must make $5. So, for horse $A$, the equation becomes

$$5 = 20 + 30 - 50X(A), \quad \text{or } X(A) = \frac{45}{50} = \frac{9}{10},$$

where the stated odds are 9:10 (that is, a winner earns 90 cents for each dollar bet).[6] Similarly,

$$X(B) = \frac{65}{30} = \frac{13}{6}, \quad \text{and } X(C) = \frac{75}{20} = \frac{15}{4}.$$

Therefore Zhihong would state the odds for $A$ of 9:10, for $B$ of 13:6, and for $C$ of 15:4.

An alternative way to compute these values is to let $p_K$ be the implied probability that horse $K = A, B, C$ will win as determined by the volume of betting. The $50 of the total of $100 bet on horse $A$ reflects the implied belief (based on the money wagered) that $p_A = 50/100$. All three values are

---

[5]When a bet is made, don't we see the odds? Yes, but they are based only on how much money has been bet on each horse up to the moment; the final odds are determined by the money bet when the betting stops. This feature is captured in movies where heavy bets are made just prior to race time.

[6]Notice how these expressions assume the form of Equation 1.27.

$$p_A = 0.5, \quad p_B = 0.3, \quad p_C = 0.2.$$

Dividing each expression in Equation 1.30 by the total amount wagered, \$100, and setting it equal to the desired earning per dollar wagered leads to the general expression

$$(1 - p_K) - p_K X(K) = 0.05, \quad K = A, B, C,$$

where the right-hand side represents the objective of earning 5 cents on each dollar wagered.

Solving these equations leads to the values

| Horse | Wagered | $p_X$ | Payoff | Odds | Implied Winning Prob |
|-------|---------|-------|--------|------|----------------------|
| A | 50 | 0.5 | 0.90 | $9:10$ | $10/19 \approx 0.526$ |
| B | 30 | 0.3 | 2.16 | $6.5:3$ | $30/95 \approx 0.316$ |
| C | 20 | 0.2 | 3.75 | $3.75:1$ | $4/19 \approx 0.211$ |

(1.31)

The third column has the victory probabilities as determined by the amount of money wagered; the sum is, as it should be, unity. The fourth column has the payoff for each dollar bet. (To understand the payoff of \$0.90 on a \$1 bet, recall that this column represents the "winnings." In a horse race, as distinct from the earlier example, the establishment is not confident that betters who lose will eagerly show up to pay the losses. So, all bets are collected in advance. This means that for each dollar bet on $A$, the winning gambler receives the dollar back plus \$0.90.) The fifth column describes the payoff in terms of the "odds" that are actually used, which is in the manner of the football example.

To understand a gambler's chances in winning, recall that the likelihood *some* horse will win is unity, so the sum of the probabilities of each horse winning is unity. Adding the probabilities in the third column does, indeed, have the value 1. But now compute the "implied" odds that some horse will win as determined by the "implied probabilities" of the last column. To remind you of the computations, let $p_A(w)$, $p_A(l)$ be the implied probabilities of horse $A$ winning and losing if the odds in the fifth column represented a fair bet. If this were a fair bet, we would have

$$p_A(w)(0.90) + p_A(l)(-1.00) = 0.$$

As $p_A(l) = 1 - p_A(w)$, it follows that $p_A(w) = 10/19 \approx 0.526$.

In reality, one horse must win. If this were a fair bet, we would have $p_A(w) + p_B(w) + p_C(w) = 1$. However, this is not a fair situation; nor was it ever intended to be. In order for the track to make money, the odds must be stacked in favor of the house. To see how much by using the above numbers, we have that

$$p_A(w) + p_B(w) + p_C(w) = \frac{10}{19} + \frac{30}{95} + \frac{4}{19} \approx 1.053.$$

Clearly this game is stacked *against* the gambler. It must be if Zhihong is to ensure a profit. It is reasonable to wonder whether the odds implied in this example are outrageous. Not really. For comparison, check the odds offered on many gambles such as on sport teams; they often sum to at least 2 rather than the value of unity for a "fair" bet. (This reality provides a sense of how badly the odds are stacked against a gambler.)

## 1.4   Exercises

1. Weather forecaster Gonnarain is so certain it will rain on Sunday that he would offer a bet of 3:2 that it will happen. Remember, if it rains after someone bets with Gonnarain, Gonnarain keeps all of the money. If it does not rain, then Gonnarain pays the better $3 for each $2 that was wagered.

   On a different channel, Alwaysunny is confident that on Sunday it will not rain; she offers 4:3 odds.

   (a) A person has $100. Set up a hedge where the person will win the same amount of money no matter what happens on Sunday.
   (b) Draw the two profit lines.
   (c) How much money is the person guaranteed to win no matter what happens?
   (d) Both forecasters believe they are making a fair bet; that is, they are willing to take either side. Determine the likelihood each has for rain on Sunday, for no rain on Sunday.
   (e) Add the probabilities of rain from Alwaysunny and of no rain from Gonnarain. Add the probabilities of rain from Gonnarain and no rain from Alwaysunny. What do you notice? How does the answer help you understand when a bet is to your advantage?

2. Two people have different opinions on the Tennessee and Florida game to be played tonight. One person supports Tennessee and is willing to offer 11 to 10 odds; the other person supports Florida and is willing to offer 5 to 3 odds. Go through the analysis to determine how much to bet with each person. If Heili has $100 to invest, how much is she assured of earning?

3. In the above discussion, it is shown how to convert specified odds from a "fair bet" into the gambler's belief about the likelihood of an event happening. The following are related.

   (a) Torik gives 5:3 odds that someone will walk in late for class tomorrow. What probability does he assign for this event?
   (b) Mikko believes there is a 60% chance that at least five students from this class will be at the next basketball game. If he were to set up odds, what would they be?
   (c) Change the 60% to 75%. Now what would be the odds?

4. Suppose John offers what he believes to be a fair bet as to whether an instructor will or will not give a class quiz tomorrow. He believes there will be a quiz and offers $a{:}b$ odds. John views this as a fair bet; compute the probability he assigns to a quiz being given and a quiz not being given.

5. Returning to a possible class quiz, Sue believes there will be a quiz and offers 3:2 odds, Torik does not and offers 3:1 odds. The odds Torik is offering are enticing, and you have $200 where you can afford to lose up to $20. Now, one possibility is to bet $20 only with Torik; if there is not a quiz, he wins and keeps your $20. If there is quiz, you win and he owes you $60. But, by hedging, you can bet more without incurring any greater risk. Explain what to do and how much you can earn if there is a quiz.

6. In the Olympic finals, only three teams are represented; USA, England, and China. Svetlana believes the USA will win and offers even odds of 1:1. Roberto is supporting England and is offering 2:1 odds. (So, if England wins, Roberto keeps his money. If one of the other teams wins, Roberto pays $2 for each dollar bet.) Finally, Jeff supports China as given by his 5:4 odds. Determine the optimal way of dividing $100 to bet that ensures the largest winning.

    A message from this example is that whenever exuberance causes individuals to offer strong odds, such as during a political campaign, arbitrage is available. To see the arbitrage opportunities from this example, if a third of the money is bet with each person, only one can win; the winnings from one other bet cover the loss and winnings from the third are profit.

    But this exercise, as with the gender example, also emphasizes the importance of understanding the provided information. When betting with one of these individuals, such as Jeff, is the bet whether China will or will not win, or does the bet require you to put forth a team and you win money only if your team wins. The difference is huge; the intent is the first, but, as a challenge, can the second be handled?

7. With a Swedish Krona, suppose the likelihood of getting *H*eads when it is spun on edge is 0.2. If X is the random variable where

$$X(H) = 1, \quad X(T) = -1,$$

find the expected value $E(X)$, the variance, $\mathrm{Var}(X)$, and express X in its standard form.

8. With an American penny, the likelihood of getting H when it is spun on edge is 0.3. If $X$ is the random variable where $X(H) = 1$, $X(T) = -1$, find the expected value E(X), the variance, $\mathrm{Var}(X)$, and express X in its standard form.

9. The next problem involves dice.

    (a) In rolling a die, it is called a "fair die" if each of the six number are equally likely, which means that the probability of a particular number appearing is 1/6. What is the expected value when rolling a die? The variance?

    (b) When rolling two fair dice and adding the numbers on the faces, the number 1 cannot occur, so this event has probability 0. What is the probability of 2, of 3, of 4, of ...? (Hint: In doing this problem, assume you have a red and a green die.) What is the expected value? Variance?

10. Suppose the probability of selecting a number from the interval $[0, 2]$ is given by $f(x) = Cx$.

    (a) Find the value of $C$ so that this is a probability distribution.
    (b) Find the probability that the given number is in the interval $[0,1]$. Now find the probability that the selected number is in the interval $[1, 2]$. The answers differ. Why?
    (c) If $X(x) = x$ is the random variable giving the value of the selected number, what is $E(X)$? The $\mathrm{Var}(X)$?

11. Suppose the PDF is given by $f(x)$, which is zero for $x < 0$ and $x > 4$. It has a constant value for $0 \leq x \leq 4$. (This is a uniform distribution; each point is equally likely.)

    (a) Find the constant value.
    (b) Find the likelihood that when selecting a point at random with this distribution, it is greater than 3.5.
    (c) If $X(x) = x$ (so, it equals the selected value), find $E(X)$ and $\mathrm{Var}(X)$.

12. Suppose the PDF is given by $f(x) = Cx^2$ for $1 \leq x \leq 2$, and zero otherwise.

    (a) Find the value of $C$.
    (b) Find the likelihood that a point selected at random with respect to this PDF is between 1.5 and 2.
    (c) With $X(x) = x$, find $E(X)$ and $\mathrm{Var}(X)$.

13. A company that prints Blue Books for exams makes a profit according to the number of books sold. Suppose that the Profit is

$$P(x) = 2(1 - e^{-2x}),$$

where $x$ is the demand or number of books sold. The demand, of course, is measured in terms of a probability distribution. After all, we cannot be assured that at any time there will be a specified demand. So, the PDF for the demand is given by

$$f(x) = \begin{cases} 6e^{-6x} & \text{for } x > 0, \text{ and} \\ 0 & \text{for } x < 0. \end{cases}$$

    (a) Show that the above is a PDF.
    (b) Find the company's expected profit.
    (c) The company worries about the variation in the profit. So, set up the integral to find the variance of the profit.

14. Show that $\text{Var}(aX + b) = a^2\text{Var}(X)$.

    We know that the standard form is $Z = \frac{X - E(X)}{\sigma}$ where $\sigma^2 = \text{Var}(X)$. Use the above to compute the $\text{Var}(Z)$.

15. The fact that $\text{Var}(X) = E(X^2) - [E(X)]^2$ was proved using properties of integrals. Prove it for the discrete case.

16. Suppose at an off-track betting facility, just before the four horse race starts, it is learned that 600 dollar bets have been made on A, 300 on B, 400 on C, and 200 on D. In order for the owner to earn 3 cents on each dollar bet, determine the payoff, or winnings, for each horse.

17. In the first test of a course, Barb got 66 and Dave got 71, where the average was 70 and $\sigma = 25$, while on the second test Barb earned 87 and Dave received 84, the average was 85 and $\sigma = 5$. They are close, but show why it is arguable that Barb did better over the two tests.

# Chapter 2
# Options

A valued lesson from Chapter 1 is that a way to reduce risk is to bet on both sides of an issue. To apply this theme to finance, appropriate market devices, with features similar what has been discussed, are identified.

## 2.1 Calls

Anna buys oranges from a supplier in order to sell them to grocery stores. To survive in this business, stability is required. The fate of our poor distributor is influenced by weather conditions, insects, economic conditions, trade wars, or whatever events may, or may not conspire to drive Anna into bankruptcy. For a multitude of reasons, it may be in Anna's best interest to sign a contract to ensure that on a specified date the correct amount of oranges *can be* purchased from the supplier at an agreed upon price.

The object, then, is to find a supplier and sign a contract for a *Call.* This means that the distributor *can,* if she wishes, *call* the supplier on the specified date, designated as the *expiration date T*, for a specified number of oranges at a specified price that is called the *exercise* or *strike price E*. Key to this description is the word *can*; the distributor *can* enforce this contract if she wishes,[1] but she need not. The contract provides her a legal right, not an obligation. The contract, however, obligates the supplier.

---

[1] The word "can" is critical. In contrast, a "futures contract" (which is an interesting topic not covered here) loses this flexibility; it requires buying or selling at a specified time and price. To illustrate a difference, with decreasing prices, there is no reason to exercise a Call, but a futures contract remains fixed. This explains why, for instance, there can be a time delay between sharply decreasing oil prices and lower gasoline prices—the refineries are locked in to the higher future contract prices.

© Springer Nature Switzerland AG 2019

D. G. Saari, *Mathematics of Finance*, Undergraduate Texts in Mathematics,
https://doi.org/10.1007/978-3-030-25443-8_2

### 2.1.1  A Long Call

To make these notions more precise, suppose there will be a rare book convention on March 9. Katri has an first edition of Agatha Christie's "Murder on the Orient Express" that is in excellent condition; today, she could sell it for $100.[2]  Erik is interested, but he is not sure whether to buy it now or at the convention. On the other hand, Erik worries whether the book price will increase to, say, $120 by convention time. The problem confronting Erik is to decide whether

1. to buy the book now, or
2. to secure a legal agreement from Katri giving him the option of buying the book on the day of the convention for $100. (So, the expiration date is $T$ = March 9, with strike price $E$ =$100.)

Again, with the Call, Erik—the purchaser—has the *legal right to decide what will be done.* If Katri agrees to the contract, she has an obligation; if he *calls* to purchase the book, she *must* go along with Erik's decision.


**Profit from a Call**

To determine the value of this contract for Erik, remember, a Call provides a legal right—whether to buy, or not to buy, the circumstance is the question. Erik's decision depends on what happens on the day of the convention—the expiration date.

If on March 9 the going price of a "Murder on the Orient Express" book is less than $100, Erik has no interest in buying Katri's copy. It would be foolish for him to exercise the agreement when he can purchase the book elsewhere at a lower price. On the other hand, should the book value be $120, then Erik would enforce the contract because, by buying Katri's book at the agreed strike price of $100, he makes a profit of $20.

The Figure 2.1 solid line describes this situation. Erik's "profit" (savings, etc.) at time $t = T$ should the book become more expensive is the difference between the current price $S$ and the strike price $E$, or $S - E$. In general, Erik's profit is

$$\text{Profit from Call}_E(S, T) = \max(0, S - E) \tag{2.1}$$

as plotted with the profit line in Figure 2.1 with a slope of either 0 or $+1$.

---

[2]There is a wide range of prices for rare books. A pristine copy of this novel, complete with book cover, would sell in the thousands. More commonly found copies in excellent shape would be in the range of the example.

**Fig. 2.1** Analyzing a Call

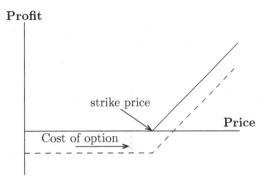

## How Much?

Why would Katri, a brilliant woman, agree to such a silly contract? After all, this Call cedes the advantage to Erik, while Katri incurs an obligation. Erik can decide what will happen—whether or not to exercise the option—and it is in his best interest to buy the book only in circumstances that hurt Katri.

Katri would endow Erik with this power *only for a price.* For a certain amount of money, she would sign the agreement. The problem is to determine the value of the contract. How much should she charge?

This pricing issue is a topic that will be analyzed in great detail after developing the appropriate mathematical tools. To develop intuition by examining a simple illustration, suppose it is accepted that there is

- a $\frac{1}{3}$ chance the value of the book will reach $120, and

- a $\frac{2}{3}$ chance the value will drop to $90.

Maybe she should charge Erik the difference between the "Expected value of the book on March 9" and the strike price.

To analyze this strategy, the expected value is

$$E(\text{book value}) = \frac{1}{3}(120) + \frac{2}{3}(90) = 100.$$

Because the difference between the expected value and the strike price is zero, this pricing approach suggests that Katri would charge Erik *nothing* to give him a right. How reckless! She most surely will not do that!

An alternative way to determine the value of the option is to emphasize what it means to Erik: Base the price on Erik's expected earnings with the Call. This approach shifts the emphasis to Erik's potential profits. Namely:

- With probability 1/3, the price will be $120. If this is the case, Erik will purchase the book and quickly resell it to obtain a profit of $20.
- With probability 2/3, the price is less than or equal to $100. Here Erik will not buy the book from Katri for $100 because he would take a loss. In this setting, Erik's earnings are $0.

The random variable from this approach is the expected profit resulting from the Call, which is

$$E(\text{Call profit}) = \frac{1}{3}[120 - 100] + \frac{2}{3}0 = 6\frac{2}{3},$$

and which could be the option's value. With this information, the price of the Call becomes a "fair bet;" neither Katri nor Erik would have an unfair advantage over the other.

Be warned; this is *not* how the value of the Call is computed. This example, for instance, specifies probabilities that most surely are never known in practice. As an illustration, what is the probability of Google stock jumping precisely $10 tomorrow? Secondly, Calls are bought and sold to make money, so, as we will discover in Chapter 6 with the Black–Scholes equation for the pricing of a Call, market pressures and arbitrage effects are involved. Nevertheless, a virtue of this discussion is that it captures the complexity of the problem while suggesting what can cause a Call's price to vary.

To appreciate how the strike price affects the option's value, the same computations with the larger strike price of $110 lead to the lower

$$\frac{1}{3}[120 - 110] + \frac{2}{3}0 = 3\frac{1}{3}.$$

This makes sense; a higher strike price offers less profit, so the Call has a smaller value. Conversely, a smaller strike price leads to a larger expected profit, so the value of the Call increases. In general,

*expect a Call's value to monotonically decrease with an increase in the strike price.*

Suppose Erik invests the $6\frac{2}{3}$ to buy the option at the $100 strike price. Once this is done, the Figure 2.1 profit line must be translated downwards by this value, which is given by the Figure 2.1 dashed lines. So, for Erik to break even, the price must reach at least $106\frac{2}{3}$. With a higher price, Erik makes a profit. Should the price reach $120, Erik reaps a $[20 - 6\frac{2}{3}]/(6\frac{2}{3}) = 200\%$ profit. If the price goes down, Erik would not buy the book, where he would lose the option price causing a -100% loss.

A question left to the reader: What should Erik do at $t = T$ (expiration date) if the price is $105, which is lower than needed to make a profit?

**Comparison**

Is this "Call" a wise idea? Such a question can only be answered by comparing alternatives. Suppose, for example, that Tyler deals in old books, and he has $100 to invest. Tyler could buy one copy of "Murder on the Orient Express" now, or he could buy as many Calls as he can afford.

To make the numbers come out right, suppose it costs $5 to purchase a Call with strike price $100. At this charge, Tyler could purchase $100/5 = 20$ Calls.

- If the price jumps to $120, each Call would be exercised. Once each copy of "Murder on the Orient Express" is bought, it is instantly sold at the prevailing price leading to a profit of $120 - 100 = $20$. Take off the $5 expense of purchasing the Call, and the profit is $15. For the 20 Calls, the total profit is $300 on an investment of $100, or a 300% profit.
- If the price drops to $90, no Call is exercised. All money is lost.

On the other hand, if Tyler is risk-adverse and purchased the book, he would have the following scenarios:

- If the price jumps to $120, he could sell it leading to a $20 profit.
- If the price drops to $90, the loss is 10% or $10.

Different approaches offer different levels of risk and return.

As a summary, buying the book now ties up money with smaller percentage of profits or losses—but with a smaller risk of going broke. By purchasing an option,

1. the buyer invests smaller amounts of money
2. with a larger risk of going broke,
3. with the potential advantage of sizable profits. Indeed, there is no limit to the amount of profit. This occurs, however, only if the price increases beyond the strike price.

## 2.1.2 Going Short

Buying the Call is referred to as taking a *long* position, which is denoted here by $C_E(S, t)$ where $E$ is the strike price and $S$ refers both to the commodity (often stock) and its price. The "$t$" variable is time, which reflects the reality that the value of the Call will change with events over time.

Selling a Call is referred to as taking a *short position;* here a negative sign captures the short position, or $-C_E(S, t)$. In the "Murder on the Orient Express" illustration, Heili sold the option to Tyler, so she is going short. Her profit curve is given in Figure 2.2, which demonstrates that Heili's profit derives from selling the option. From this curve, the advantages and disadvantages can be determined.

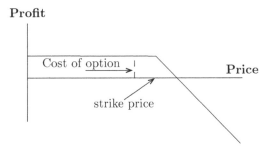

**Fig. 2.2** The analysis of a short Call

1. Going short on a Call has unbounded risk; if the price increases to sufficiently high levels, the loss can be unbounded.
2. If the price does not increase, a guaranteed earning is obtained.

As with the football game, different people can have different opinions about what will happen—here it is how prices might change. The person with a long Call expects the prices to increase; the person with a short Call believes the prices will remain the same or decrease.

To offer another example of taking the short position, suppose John wants to go short on Mesmerized stock; this means he wants to sell this stock. There is a slight complication; John *doesn't own any*. No problem: The way this is accomplished, with appropriate controls to avoid abuse, is that John calls his broker and asks to go short. The broker "borrows" the stock from somewhere, maybe from Mary, sells it, and gives the money to John. The notation used here for going short on the stock is $-S$.

But John must replace that stock at a specified time, so why would he do this? What is John's expectation about the prices?

### 2.1.3   Hedging

John just went short of the Mesmerized stock; he sold it for $100. He did so because he expects the cost of Mesmerized stock to decrease. If, for instance, it drops to $80, John can quickly buy the replacement stock for $80, and return it to Mary for a nice quick profit of $100-$80 = $20.

Fine, but what a risk! Suppose the public suddenly becomes bedazzled with Mesmerized stock making it so popular that the price jumps to $125. Poor John must buy this replacement stock for $125 to return it to Mary for a loss of $25. This is where lessons from the football gambles (Chapter 1) provide insight: Maybe there is a way to hedge.

Remember, a purpose of a hedge is to provide protection for John should the price increase. As described above, a Call is based on the assumption that the price will increase; going short on the stock is based on the assumption that the price will decrease. Combining a Call with going short, denoted by

$$C_{100}(S, t) - S,$$

constitutes betting on whatever will happen.

To appreciate this hedge, if the price decreases, as John expects, he can ignore the Call and earn money by buying a replacement stock on the market at the lower price. But should the price skyrocket to, say, $130, John is protected. Rather than buying the replacement stock at the high price of $130, he can exercise the Call to obtain the replacement for $100. All John loses is the cost of the Call.

This hedge, which combines a short sale with a Call, is a form of insurance. More about how options provide protection is described as we continue.

## 2.2  Puts

The opposite of a Call is a Put: Here a person has the *right* to sell something—to *put* it on sale—at a specified price (strike price $E$) on a specified date (expiration date $T$). It is denote by $P_E(S, t)$.

### 2.2.1  A Long Put

Adrian, who sells old books, believes that the $100 price for a copy of "Murder on the Orient Express" is overvalued; the price will drop. He doesn't have a copy, but Adrian would like to take advantage of his expectation that the price will drop by selling one to Anneli for $100 at that March 9 conference.

If the price drops, why would Anneli buy the book from Adrian for $100 when she can get it cheaper elsewhere? So, for a fee, they agree that Adrian has the *right* to decide whether or not he will sell her a book on the *expiration date* of March 9 for the *strike price* of $100. Namely, the contract Adrian purchases from Anneli—a Put denoted by $P_{100}(S, t)$—gives Adrian the right to *put* the book on the market to Anneli.

If the book price decreases, it is in Adrian's interest to exercise this option. After all, if the price drops to $90, he can buy a book from someone else, sell it to Anneli for $100, and pocket the $10 profit. On the other hand, should the price be $120, there is no way Adrian will sell Anneli the book for $100; Adrian would ignore the Put. With a Put, the benefit and rights belong to the seller.

Again, the concern is to determine how much Anneli should charge for the Put. A rough approach to compute this fee is in terms of the expected profits to the person holding the option. Using the earlier assumptions about this book:

- With probability 1/3 the book value will be $120. Here, Adrian will *not* sell the book, so his profit is zero.
- With probability 2/3, the book value will *drop* to $90. Here, Adrian will exercise the Put: His profit will be $10,

Thus, the expected value of the Put option is

$$\frac{1}{3}0 + \frac{2}{3}10 = \frac{20}{3} = 6\frac{2}{3}.$$

Therefore, the value of this option is $6\frac{2}{3}$, which happens to agree with the value of the Call with the same strike price.

This price agreement is a coincidence. To demonstrate how the values of a Put and Call can differ, suppose the strike price for the Put is $110—that is, on March 9 Anneli must buy the book from Adrian for $110 if he wishes to sell it. Clearly, with the higher strike price, Adrian can make a larger profit: The expected value of the Put now is

$$\frac{1}{3}0 + \frac{2}{3}[110 - 90] = \frac{40}{3} = 13\frac{1}{3}.$$

So,

> *with an increase in the strike price the value of a Call decreases,*
> *but the value of a Put increases.*                                    (2.2)

Returning to the Put, if the price decreases, Adrian can buy a book at the lower price to sell to Anneli, which ensures a profit for Adrian. If the price increases, Adrian will not exercise this Put option because he would lose money, and incurs only the cost of buying the option. Figure 2.3, which graphs Adrian's situation, displays the advantages and disadvantages of a Put. The profit is given by

$$P_E(S, T) = \max(0, E - S).  \qquad (2.3)$$

**Fig. 2.3** The analysis of a Put

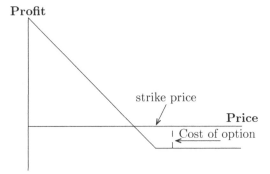

as depicted in Figure 2.3.

1. A profit is made on a long put if the price drops.
2. Limited risk.

### 2.2.2 Short Put

A "short Put" is where a person sells such a "Put" contract. The profit curve relative to changes in the price is depicted in Figure 2.4.

### 2.2.3 Some Jargon

To conclude this section, it is worth mentioning some of the market jargon. A common one is "in the money" represented by ITM. Let me ask the reader to predict what this means when applied to a Call or a Put.

As one might guess, "in the money" is when the option is, well, in the money; it is worth something. Now, $C_E(S, t)$ is worth something if $S > E$, so a Call is "in the money" if the current, or spot price, is greater than the strike price. For a Put $P_E(S, t)$, everything is reversed: Its value derives from the spot price being below $E$, so a Put is "in the money" if the spot price $S$ is below the strike price $E$.

The "out of the money," or OTM, has the obvious and opposite meaning. This OTM designation for an option is where $S$ is not causing the option to be of interest. Thus $C_E(S, t)$ is out of the money if the spot price $S$ is smaller than the strike price $E$, and $P_E(S, t)$ is out of the money if the current $S$ is greater than the strike price. What reflects the contrasting roles of Puts and Calls is that $C_E(S, t)$ is OTM when $P_E(S, t)$ is ITM.

**Fig. 2.4** The analysis of a short Put

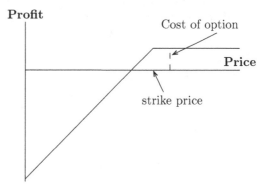

If phrases such as "in the money" and "out of the money" are tossed about, surely there is something in the middle, which would be an option "at the money," or ATM. Quite naturally, here the current price is approximately equal to the strike price.

## 2.3  Hedging

The above introduces basic terms. In all of these situations, a certain element of risk is involved. The goal, as with the football game illustration, is to discover how to minimize risk by betting on both sides.

A way to underscore the importance of hedging is to point out that many attorneys, including personal friends, have done quite well in suing money managers who did not appropriately hedge a client's investments; an oversight that caused the client to suffer unnecessary financial losses.

### 2.3.1  Straddle

With a volatile market, where it is not clear whether prices will go up or down, one strategic approach is to buy a Call and a Put for the same strike price and the same expiration date; in this manner, no matter what happens, at least one, the Call or the Put, will be in the money. Comparing this strategy with the Chapter 1 football example, it resembles betting on the Vikings and the Packers.

This portfolio

$$Port(S, t) = P_E(S, t) + C_E(S, t) \tag{2.4}$$

is called a *straddle*. In Figure 2.5, the dashed lines represent the Call and Put; the solid line is the combination, along with the cost, creating the straddle. It is left to the reader to examine the figure and determine the advantages and disadvantages.

In doing so, it is worth describing what happened years ago when I introduced this strategy to my class. At this same time, a firm was waiting for a court decision.

**Fig. 2.5**  Straddle

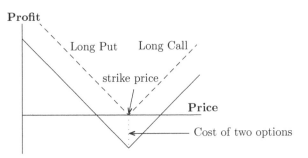

**Fig. 2.6** Strangle

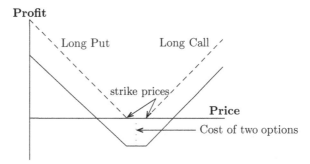

If the firm would lose, its stock value would drop; if the firm would win, the stock value would jump. And so the price would change—in some direction—but nobody knew what would happen and in which way the price would vary. The straddle handled this difficulty; if prices change, in either direction, a profit would be made. A profit would be denied only if prices remained fixed; e.g. if the court decision would be delayed. It was not, and several adventurous students benefited.

The strike prices need not be the same for the Put and Call; when they differ, the combination is called a *strangle*, which is represented in Figure 2.6. A question left to the reader is to explain why someone would prefer a strangle to a straddle. (Hint, how does a change in the strike price alter the return—and cost—of the option?)

As a related issue, the reader should explore how to use short Calls and Puts. The exercises provide examples.

### 2.3.2  Designing Portfolios

Treat the Calls and Puts as building blocks to assemble a portion of a portfolio or an investment strategy. Names for certain of the better known approaches, such as the straddle and strangle, are in Exercise 4 at the end of this chapter. But, to develop intuition, to better understand how and why different methods work, to be able to create your own strategy to fit a current opportunity, it is worth understanding how to generate new approaches. In particular, a reader should be able to take a portfolio and graph its profit curve at expiration date. Conversely, you should be able to take a desired profit curve and design an associated portfolio. Let's go after it.

An introduction comes from Figure 2.7, which captures the earlier story where John goes short on Mesmerized stock. Figure 2.7a is the profit curve. As long as the price of the stock is below $E = 100$, John will make money. This feature, where each dollar the price drops is another dollar of profit, is given by the graph above the $x$-axis with the slope of $-1$. But should the price increase above $E$, each dollar increase is a dollar loss for John, as reflected by the portion of the line below the $x$-axis.

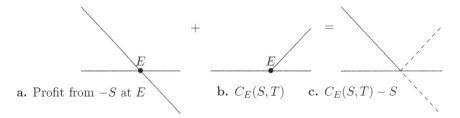

a. Profit from $-S$ at $E$          b. $C_E(S,T)$       c. $C_E(S,T) - S$

**Fig. 2.7** Going Short with a Call

**Fig. 2.8** Example profit line

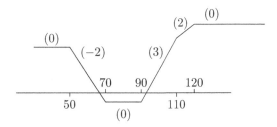

Figure 2.7b shows how the Call can come to the rescue! The Call is worth nothing for John should the price of Mesmerized stock, $S$, be below $E$, but for each dollar $S$ is above $E$, it provides a dollar profit for John, as reflected by the $C_E$ slope of $+1$. The dollar profit offsets the dollar loss John incurs by going short on the stock, so the loss and profit cancel. This leads to Figure 2.7c, which is the profit curve (without the cost of options) for $C_E(S, T) - S$. The dashed lines in Figure 2.7c represent the offsetting profit and loss.

Figure 2.8 is a practice curve that is divorced from reality. It represents a hypothetical profit curve where the goal is to determine an associated portfolio. The numbers in parenthesis, such as $(-2)$, represent the slope.

To interpret the curve, this person appears to believe the price might increase above 90, probably taper off between 110 and 120, but most surely it will not exceed 120. Although the price might decrease below 70, it probably will not sink below 50. (Be sure to understand why these comments may, or may not, reflect the shape of the curve.) Your goal is to design a portfolio; for now, ignore the cost of options, just use Puts and Calls to create the shape. Do it in three ways:

1. Create a portfolio using only Calls—long and short.
2. Create a portfolio using only Puts.
3. Today's price is 80. Create a cost-effective portfolio using a mixture of Puts and Calls.

### Using Calls

With Calls, the curve's building blocks come from Figures 2.1 and 2.2. Starting from the extreme left, the horizontal slope changes to $-2$ at $S = 50$, which requires

(Figure 2.2) going short: The first term is $-2C_{50}(S, t)$. At $S = 70$, the slope becomes zero. Now, each dollar beyond $S = 70$ means that $-2C_{50}(S, t)$ is losing two dollars. A way to earn a dollar for each dollar of price increase above $S = 70$ is with a Call. Thus, to offset the losses from $-2C_{50}(S, t)$, purchase two Calls leading to $-2C_{50}(S, t) + 2C_{70}(S, t)$.

The next change at $S = 90$ is where the slope jumps to 3 to earn three dollars for each dollar $S$ goes above 90; this means we need three Calls for $-2C_{50}(S, t) + 2C_{70}(S, t) + 3C_{90}(S, t)$. This portfolio provides three dollars for each dollar $S$ moves beyond 90. But the curve is tempered after $S = 110$ to receive two dollars for each dollar increase. This decrease from 3 to 2 is achieved by going short to have

$$-2C_{50}(S, t) + 2C_{70}(S, t) + 3C_{90}(S, t) - C_{110}(S, t).$$

Finally, the objective is for no money to come in after $S = 120$. This requires going short on two Calls to obtain

$$Port = -2C_{50}(S, t) + 2C_{70}(S, t) + 3C_{90}(S, t) - C_{110}(S, t) - 2C_{120}(S, t). \quad (2.5)$$

Here is an interesting question: Why would anyone want to decrease the amount of money coming in by going short? An answer is that this person does not believe the market will reach these regions, but she can make money by going short—by selling Calls.

**Using Puts**

To create a portfolio by using Puts, the building blocks are Figures 2.3 and 2.4. A hint in doing so:

> When dealing with Calls, start from the left and work to the right. When dealing with Puts, start from the right and work to the left.

Following this advice, when moving from the right to left, the first change is at $S = 120$ where the slope goes downward. According to Figure 2.4, go short on a Put, so the portfolio starts with $-2P_{120}(S, t)$. At $S = 110$, there is another slope that is captured by a short to give $-P_{110}(S, t) - 2P_{120}(S, t)$. At $S = 90$, the slope becomes horizontal, which requires three Puts for $3P_{90}(S, t) - P_{110}(S, t) - 2P_{120}(S, t)$. At $S = 70$, the slope increases to two, which (Figure 2.3) requires acquiring two Puts leading to $2P_{70}(S, t) + 3P_{90}(S, t) - P_{110}(S, t) - 2P_{120}(S, t)$. Finally, lower than $S = 50$, the slope is zero, which requires going short on two Puts to obtain

$$Port = -2P_{50}(S, t) + 2P_{70}(S, t) + 3P_{90}(S, t) - P_{110}(S, t) - 2P_{120}(S, t). \quad (2.6)$$

Here is a question: There are coefficient similarities between Equations 2.5 and 2.6. Can you explain why this relationship holds in general? The explanation reflects a duality of Puts and Calls.

**Mixed Choice**

For a useful combination of Puts and Calls, use Calls to the right of $S = 80$ (today's price) and Puts to the left. In doing so, follow the earlier hint of moving from the left to right with Calls. As the slope of Figure 2.8 is zero around $S = 80$, but jumps to three at 90, this portion of the portfolio is $3C_{90}(S, t)$. Proceeding as above, the Call portion is $3C_{90}(S, t) - C_{110}(S, t) - 2C_{120}(S, t)$.

Similarly, for the Put portion, start at 80 and move to the left. Thus the Put portion is $-2P_{50}(S, t) + 2P_{70}(S, t)$. Combined, the portfolio is

$$Port = -2P_{50}(S, t) + 2P_{70}(S, t) + 3C_{90}(S, t) - C_{110}(S, t) - 2C_{120}(S, t), \quad (2.7)$$

which, with the coefficients, bears a strong similarity to Equations 2.5 and 2.6. This is not a coincidence.

Which of these three possibilities, Equations 2.5, 2.6, or 2.7 should be used? Modifying a phrase introduced by James Carville during the 1992 US presidential campaign; the answer reflects the comment

*"It's the profit, stupid!"*

This is where the information that today's price of 80 enters along with the Equation 2.2 statement. The cost of Puts with a strike price above the current value can be prohibitive, while the cost of Calls is more reasonable. Conversely, the cost of Calls with a strike price below the current value can be expensive, while the cost of Puts is more reasonable. Thus, expect the choice of Equation 2.7 to be more reasonable, which leads to greater returns.

## 2.4  Put–Call Parity

Puts and Calls are financial tools. As true with any collection of instruments, a flailing, inexperienced amateur can create a mess while a craftsman can produce something beautiful. The attractive combination described in this section is a powerful tool called the "Put–Call Parity Equation," which plays a central role in the pricing of Puts and Calls.

A message resulting from this relationship is that the earlier "expected profit" analysis for pricing these tools is overly naive. This new relationship will involve interest rates and something called "The present value of money," which makes it clear that the actual pricing of Puts and Calls includes more complicated interactions between investing in a particular commodity or stock, or placing this money in a bank and collecting interest.

### 2.4.1 Present Value of Money

James has a note granting its holder $1000 on September 1. James, as always, needs money—*now*. So, he wants to sell the note to Diane. At the rate of interest of 5%, which is to be compounded continuously, what is this note worth today?[3]

Stated in another manner, Diane could buy the note from James, or she could deposit her money in the bank and collect interest. To compare her options, Diane must know how much money she needs to deposit in a bank that provides 5% interest to have precisely $1000 on the expiration date.

To set up this problem, let $M(t)$ represent money value at time $t$, $E$ is the final value of the note (for the problem, $E = \$1000$), $T$ the expiration date (which, for the problem, is September 1), $t$ is today's date, and $r$ is the interest rate. The change in money, the interest, over a very short period is approximated by the usual "interest equals rate × money × time," or

$$\Delta M(t) = r\, M(t)\, \Delta t. \tag{2.8}$$

Using the typical strategy of collecting similar variables on different sides of an equation, this expression becomes

$$\frac{\Delta M}{M} = r\, \Delta t,$$

which equates the ratio of change in money with the product of the interest rate and change in time. The goal is to sum both sides over the duration of time leading to the approximation

$$\sum \frac{\Delta M}{M} = r \sum \Delta t,$$

where the summations are over all $\Delta t$ time intervals from $t$ to $T$.

A better approximation is obtained by taking the limit of the approximating summations as the size of the $\Delta t$ time intervals shrink. Using the Equation 1.12 expression, this leads to

$$\int \frac{1}{M}\, dM = r \int dt.$$

---

[3]Many, if not most, readers are familiar with the *present value of money*, but read along because the structure leading to this expression is used in several other places.

The limits of integration are determined by the provided information. At time $t$, for instance, the note is worth (the unknown value) $M(t)$; these values determine each integral's lower limit. At time $T$, the note is worth the expressed value of $M(T) = E$. Thus the integrals become

$$\int_{M(t)}^{E} \frac{1}{M} \, dM = r \int_{t}^{T} dt.$$

As learned in a first course of calculus, the answer is

$$\ln(M)|_{M(t)}^{E} = \ln\left(\frac{E}{M(t)}\right) = r(T - t).$$

Eliminating the natural log term—by raising both sides to a power of e— leads to the *the present value of money* expression

$$M(t) = Ee^{-r(T-t)}. \tag{2.9}$$

To take an intuition break involving finance, suppose the current price of a stock is higher than the present value of $E$; that is, $S > Ee^{-r(T-t)}$. Which is worth more; $C_E(S, t)$ or $P_E(S, t)$? Is there a relationship?

## 2.4.2  Security

To appreciate how to use this information, suppose Tyler borrows money from Heili. With the interest of $r$ on the loan, Tyler promises to pay Heili $\$E$ on the expiration date $T$. As derived above, the present value of this money—the amount Tyler borrows—is given by Equation 2.9.

There is no way Heili will loan money to Tyler without some security. So, Tyler gives her his Mesmerized stock, $S$, with the expectation that this stock is worth $\$E$ on the expiration date. Right now, Heili has the stock and Tyler has the money: Both have good intensions of making an exchange on the expiration date; that is, they expect that on date $T$, Tyler will pay Heili $\$E$ and Heili will return Tyler's stock. Here is a problem; there are no legal guarantees to ensure this will happen!

Heili, a realist, does not believe in "good intensions." Consequently she seeks legal shielding to ensure that this exchange will take place. Her worry is that should the stock price drop, it is in Tyler's financial interest to keep the $\$E$ and forget about having his stock returned. So, for protection, Heili insists on a Put with strike price $\$E$ on expiration date $T$. Her reasoning is that even should the stock price crumble, the Put gives her the legal right to sell the stock back to Tyler at the price $\$E$.

The current situation, then, is that Heili has the stock, $S$, and the Put, $P_E(S, t)$; Tyler has the money worth $Ee^{-r(T-t)}$. To appreciate how this arrangement puts Tyler at a disadvantage, suppose the stock price jumps so high that at expiration date it is worth far more than $\$E$. Here, Heili would have no interest in selling the stock back to Tyler for a measly $\$E$; she can command a better price on the market. Consequently, to provide protection, Tyler insists on having a Call, which gives him the legal right to buy back the stock at the specified price of $\$E$ at time $t = T$, which is ensured through $C_E(S, t)$.

At this stage, all problems are covered with a sense of equality. Combining these expressions leads to what is known as the "Put–Call Parity Equation"

$$S + P_E(S, t) = C_E(S, t) + Ee^{-r(T-t)}. \tag{2.10}$$

Consistent with the story is that the asset ($S$) and the power (the Put) to sell it are on one side of Equation 2.10, while the power to buy an asset (the Call) and the money to do so (the $Ee^{-r(T-t)}$) are on the other side.

It will be shown that Equation 2.10 is, indeed, an equality. But first, *any* relationship should be examined to determine whether it provides useful market information. This expression, for instance, asserts that subtle market pressures, such as the interest rate $r$, affect the Call and Put prices. This is because the difference of their values at price $S$ is

$$C_E(S, t) - P_E(S, t) = S - Ee^{-r(T-t)}. \tag{2.11}$$

The answer for the question following Equation 2.9 now is immediate. If $S$ is larger than the present value of $E$ (that is, $S > Ee^{r(T-t)}$), then $C_E(S, t)$ is worth more than $P_E(S, t)$ where the *precise* difference follows from Equation 2.11.

**Example**  Today's price of Ecstatic Stock is $\$100$, the interest rate $r = 0.04$, and a Call with expiration date one-fourth of a year from now with strike price $\$110$ is worth $\$11$. How much is a Put, with the same expiration date and strike price, worth?

To compute with Equation 2.10 by substituting in the knowns, the expression becomes

$$100 + P_{110}(100, t) = 11 + 110e^{-0.04(1/4)} = 11 + 110e^{-0.01} = 11 + 108.91,$$

or

$$P_{110}(100, t) = 19.91.$$

For an intuition break, suppose all information is the same except that today's price of the stock is $\$115$ and the price of the Call is $\$2$. Is there anything wrong with these figures? Instead of $\$2$, compute a lower bound on the price for the Call with this information.

**Back to the Put–Call Parity Equation.**  To prove that Equation 2.10 always holds, first consider what happens at time $t = T$. The easy case is if $S = E$ because at this price neither the Put nor the Call will be used; i.e. $P_E(E, T) = C_E(E, T) = 0$. (See Figures 2.1, 2.3.) Consequently, Equation 2.10 becomes the obvious equality

$$E + 0 = 0 + Ee^{-r(T-T)} = E.$$

The first of two remaining cases is that $S < E$, where the term on the right-hand side of Equation 2.10 equals

$$C_E(S, T) + Ee^{-r(T-T)} = 0 + E.$$

After all, with $S < E$, the Call is worth zero (Figure 2.1).

On the left-hand side, because $S < E$, the Put will be exercised (Figure 2.3), so the asset $S$ is sold for $\$E$. Thus, this left side equals

$$S + P_E(S, T) = S + (E - S) = E.$$

Both sides of Equation 2.10 are equal.

For the remaining $S > E$ case, the Put is useless (Figure 2.3), so the left-hand side of Equation 2.10 equals

$$S + P_E(S, T) = S.$$

On the right-hand side, spend $\$E$ and exercise the Call to purchase the stock for the specified price of $\$E$; the profit from the Call is $\$(S - E)$. Thus, the right-hand side equals $C_E(S, T) + E = (S - E) + E = S$, and Equation 2.10 is satisfied.

**What Happens for $t < T$?** Credit for why Equation 2.10 holds (well, up to transaction costs, etc.) for $t < T$ belongs to our old friend "arbitrage." The reason is that if the two sides fail to agree, the disagreement introduces an opportunity to make some money—risk free. To see how to do so, suppose at time $t < T$ that, rather than equality,

$$S + P_E(S, t) > C_E(S, t) + Ee^{-r(T-t)}. \tag{2.12}$$

Remember the old phrase, "Buy low, sell high." As the "high side" is $S + P_E(S, t)$, Mikko sells it by going short. Perhaps he goes short on the Put with Torik (so Torik purchases the Put, which gives Mikko money) and borrows stock from Tatjana to sell at price $S$. In this way, Mikko has enough cash to buy the $C_E(S, t) + Ee^{-r(T-t)}$ package *and* have extra money! (The excess is the difference between the left and right sides of Equation 2.12.) "Buying" $Ee^{-r(T-t)}$ means investing this amount of money in the bank.

According to the above, at expiration date $t = T$ the Put–Call Parity Equation holds. For assets, Mikko has $\$E$ from the bank and a Call; against him is a FF and the need to replace the borrowed asset (by going "short") to Tatjana. No problem. If $S = E$, neither the Call or Put would be exercised, so Mikko can use the $\$E$ from the bank to buy the asset and return it to Tatjana.

Suppose $S > E$. Remember, Mikko is short on the Put, which allows Torik to exercise it. There are no worries, because Torik will not do so. If he did, he would "require" Mikko to buy an asset at a *reduced price* of $E$; Mikko would happily do so as he could immediately sell it for $\$S$ and a $S - E$ profit—with Torik suffering a loss. All that remains is for Mikko to replace the asset that he owes Tatjana. But Mikko has the Call $C_E(S, T)$ that he can exercise by using the $\$E$ that he has in the bank to purchase the asset and return it to Tatjana. Everything is fine; Mikko can meet his obligations at $t = T$—and he *still has that excess arbitrage profit!*

Similarly, if $S < E$, the Call Mikko purchased is useless, so all Mikko has is the $\$E$ from the bank. But, Mikko is short on a Put: Because $S < E$, Torik would exercise the Put to force Mikko to buy the asset at the higher price of $\$E$! This is fine with Mikko; he needed to buy the asset anyway in order to return it to Tatjana. All obligations are cleared, and Mikko still has the added arbitrage profit.

With an available arbitrage opportunity, wise people will jump in. With many selling Puts and buying Calls, the demand for Puts goes down, so the price of a Put will decrease. Similarly, with the increased demand to buy Calls, the cost of a Call will go up. It will continue doing so until the arbitrage opportunities are squeezed out of the system, which is the "Put–Call Parity" relationship.

The argument on the other side of the equation is similar. Namely, if it is true at $t < T$ that

$$S + P_E(S, t) < C_E(S, t) + Ee^{-r(T-t)}, \tag{2.13}$$

then "sell high." That is, sell $C_E(S, t) + Ee^{-r(T-t)}$, which means Mikko is going short on the Call and borrowing $\$Ee^{-r(T-t)}$ from the bank. With the money earned from the Call and borrowed from the bank, Mikko can buy $S + P_E(S, t)$. The money left over is his arbitrage profit.

At time $t = T$, accounts must be settled. Remember, Mikko purchased $S + P_E(S, T)$, with debts coming from $C_E(S, T) + E$; more precisely, he must pay the bank $\$E$. With a similar argument, if $S > E$, then Torik, who has the Call, will call Mikko to buy his asset for the reduced price of $\$E$. Fine; Mikko now has the money to pay off the bank. Similarly, should $S < E$, then Mikko can exercise his Put to sell the asset for $\$E$ and pay off the bank.

## 2.5 Information Gained

The Put–Call Parity Equation provides insightful information concerning the value of a Put and a Call. Start with $C_E(S, t)$. As described above, at time $t = T$ the value of

$$C(S, T) = \max(S - E, 0). \tag{2.14}$$

What happens for $t < T$?

The Put–Call Parity Equation asserts that at $t < T$, the $E$ value is replaced by the present value of $E$ (as given with the bracketed term in Equation 2.15), with a buffer supplied by the price of a Put. In other words, replace Equation 2.14 with the more general

$$C_E(S, t) = [S - Ee^{-r(T-t)}] + P_E(S, t). \tag{2.15}$$

At time $t = T$, the value of Equation 2.15 reverts to Equation 2.14. (Carry out the details to show this.)

A similar expression replaces the $P_E(S, T) = \max(0, E - S)$ with the more general assertion

$$P_E(S, t) = [Ee^{-r(T-t)} - S] + C_E(S, t). \tag{2.16}$$

that holds for $t \leq T$ Again, Equation 2.16 reverts to the standard value when $t = T$. Also, at the beginning of this chapter, a relationship between the price of, say, a Call and the expected profit was explored. But the actual price is determined by what agents are willing to pay; it is determined by market pressures. What we learn from the Put–Call Parity Equation is that *arbitrage* plays a strong role in setting these values; that is, *the price of a Call or Put is an arbitrage value*. This theme is further explored next.

### 2.5.1   Our Friend: Arbitrage

To explore what else can be squeezed out of this Put–Call expression, suppose all that is known for $T - t$ equal to one year and $r = 0.02$ is that

$$C_{50}(51, t) - P_{50}(51, t) = 7.$$

How can this information be used, or exploited?

To appreciate the use of this information, rewrite the Put–Call equation as

$$C_E(S, t) - P_E(S, t) = [S - Ee^{-r(T-t)}], \tag{2.17}$$

which means that the difference in cost between a Call and a Put should be the difference between current price $S$ and the present value of $E$. But in this particular case,

$$[S - Ee^{-r(T-t)}] = 51 - 50e^{-0.02(1)} \approx 2, \tag{2.18}$$

which creates a large discrepancy! So, "sell high, buy low" by going short on the $C_{50}(51, t) - P_{50}(51, t)$. That is, Torik, who follows this strategy, sells the Call, buys the Put to acquire $P_{50}(S, t) - C_{50}(S, t)$ *plus* \$7. The next step is for Torik to buy low, so he purchases $[S - Ee^{-r(T-t)}]$.

A problem for Torik is to find where to get the money. Part from the bank because, remember, the $-50e^{-r(T-t)}$ terms means that Torik borrows $\$50e^{-0.02}$. The rest comes from the \$7 Torik earned by going short on $C_{50}(51, t) - P_{50}(51, t)$. Thus (from Equation 2.18), after everything is bought and sold, the extra,

$$7 + 50e^{-0.02} - 51 \approx 7 - 2 = 5,$$

goes into Torik's pocket as arbitrage earnings. In addition, Torik has a Put, he is short on a Call, and he has the item—*but* he must repay \$50.

On expiration date, that \$50 loan must be repaid, which, somehow, must come from Torik's $P_{50}(S, T) - C_{50}(S, T) + S$ holdings. To see how this is done, if $S > 50$, then the Put is worth nothing. But Torik went short on a Call, so the person owning the Call will call Torik to make a profit by buying Torik's asset $S$ for \$50. Fine; Torik now has the \$50 to replay his loan.

If $S < 50$, the Call is worth nothing (Figure 2.1), but the Put allows Torik to sell the item for \$50, and repay the loan. Of course, if $S = \$50$, just sell it and pay off the loan. Important is that these transactions do not involve the arbitrage profit of \$5 that Torik earned by selling high and buying low.

Again, our friend "arbitrage" kicks in; this is where observant people on the market buy and sell appropriate items from the Put–Call equation to realize a profit. With many people wishing to go short on $C_{50}(51, t) - P_{50}(51, t)$, the cost of this combination will decrease causing arbitrage opportunities to fade away. This analysis reaffirms an earlier comment that, while principles of physics allow making mathematical assumptions of continuity, what permits such assumptions in finance is the mathematical friend arbitrage. If Equation 2.10 does not hold, opportunities for arbitrage do!

## 2.5.2 Properties of Puts and Calls

The Put–Call Parity Equation introduces refinements on our understanding of the cost of Puts and Calls. As argued earlier, for instance, if $E_1 < E_2$, then anticipate the cost of $C_{E_1}(S, t)$ to be greater than $C_{E_2}(S, t)$. That is, a higher strike price decreases the Call's value. To appreciate why this is so, at expiration date and if $S$ is large enough, the profit of $C_{E_j}(S, T) = S - E_j$. Thus the smaller the $E_j$ the larger the profit, or that $C_{E_1}$ is potentially more valuable than $C_{E_2}$. And so, the cost should be more.

Similarly, if $S > E_1, E_2$, the lower bounds for $C_{E_j}(S, t)$ coming from the Put–Call Parity Equation are

$$C_{E_1}(S, t) > S - E_1 e^{-r(T-t)} \text{ and } C_{E_2}(S, t) > S - E_2 e^{-r(T-t)}, \tag{2.19}$$

or the difference between $S$ and the present value of $E_j$. So, the smaller $E_j$ value creates a larger lower bound, which confirms suspicions. A similar analysis applies to Puts, except that the costs are reversed; that is, $P_{E_2}(S, t) \geq P_{E_1}(S, t)$, which reflects the sense that with $E_2$ being larger, profits start coming in at an earlier $S$ value.

### Interest and Other Rates

Everywhere in these notes, it is assumed that the interest rate $r$ not only is fixed but is the same for money in the bank and money being borrowed from the bank. This is not true! But it is a standard simplification.

If $r$ increases, how does this affect the cost of Puts and Calls? Again, help comes from versions of our new friend the Put–Call Parity Equation. To see how it can be used, the Equation 2.11 version

$$C_E(S, t) - P_E(S, t) = S - Ee^{-r(T-t)}$$

states that the difference between cost of the two options is determined by the current difference between the price of $S$ and the present value of $E$.

An increase in $r$ means that a smaller amount of money needs to be deposited in the bank to receive $\$E$ on the scheduled date. In turn, the present value of $E$ is smaller. So, if $S > E$, it follows that:

- With an increase in $r$, the $C_E(S, t) - P_E(S, t)$ difference increases.
- With an increase in $E$, the $C_E(S, t) - P_E(S, t)$ difference decreases.
- With an increase in $S$, the $C_E(S, t) - P_E(S, t)$ difference increases.
- With an increase in $T - t$ (that is, more time until expiration date), $C_E(S, t) - P_E(S, t)$ difference increases.

Of course, if $S$ is less than the present value of $E$, then an analysis is simplified by rewriting the expression as

$$P_E(S, t) - C_E(S, t) = Ee^{r(T-t)} - S,$$

and a similar analysis applies.

## 2.6  Exercises

1. Suppose on the strike date, the price of a security will be $120 with probability $\frac{1}{4}$, $100 with probability $\frac{1}{2}$, and $80 with probability $\frac{1}{4}$.

   (a) Use the crude, expected profit approach to figure the value of a call option with strike price 100.
   (b) Do the same for strike price 110.
   (c) Do the same for a put with the above two strike prices.
   (d) What basic points become clear from this exercise?

2. Instead of using the expected profit approach, try a "fair bet" approach. The idea is that if a dealer charges too much for a Call, people will not buy it. If the dealer charge too little, she runs a risk. To eliminate the risk, treat this as a "fair bet." Namely, charge $\$x$ for a Call where $x$ is the value that makes my expected winnings equal to zero. Compute this cost with the above data for a Call with strike price $\$100$. Explain any similarities or differences with the answer from the first part of Exercise 1.

3. Sandra bought item $S$ for $\$65$ with the intent of selling it to Sam on March 1. If the price for the item jumps to $\$100$, which is her guess, she will make $\$100 - \$65 = \$35$.

   (a) Draw the profit curve.
   (b) If the price drops to $\$20$, she is incurring a $\$20 - \$65 = -\$45$ loss. (This should be in the profit curve.) What can she do to protect herself? What should be her portfolio?

4. What follows are several portfolios; each is denoted by the term $Port$. Graph the profit curve at $t = T$ for each of them. Determine the strengths and weaknesses of each portfolio. Namely, explain what a reasonable investor must be expecting to adopt each portfolio.

   In this listing, $C_E$ denotes a Call with strike price $E$, $-C_E$ represents going short on a Call—the Call was sold, $P_E$ is a put, $-P_E$ is going short on a put, and the numbers $E$ indicate the strike price.

   (a) A *"Long Call"* is $Port = C_E$.
   (b) A *"Short Call"* is $Port = -C_E$.
   (c) A *"Long Put"* is $Port = P_E$.
   (d) A *"Short Put"* is $Port = -P_E$.
   (e) A *"Long Straddle"* is $Port = P_E + C_E$; namely, buy both a Put and Call with the same strike price.
   (f) A *"Short Straddle"* is $Port = -P - C$.
   (g) A *"Long Strangle"* is $Port = P_{E_1} + C_{E_2}$; namely, it is a purchase of a put at strike price $E_1$, which is less than the strike price of a Call at $E_2$
   (h) a *"Short butterfly"* is $Port = -C_{E_1} + 2C_{E_2} - C_{E_3}$ where $E_1 < E_2 < E_3$.
   (i) A *"Long butterfly"* is $Port = C_{E_1} - 2C_{E_2} + C_{E_3})$ where $E_1 < E_2 < E_3$.
   (j) A *"Short Condor"* is $Port = -C_{E_1} + C_{E_2} + C_{E_3} - C_{E_4}$ where $E_1 < E_2 < E_3 < E_4$.
   (k) A *"Long Condor"* is $Port = C_{E_1} - C_{E_2} - C_{E_3} + C_{E_4}$

5. Compare a long straddle with a long strangle explaining advantages of each.
6. This is another question about portfolios. Here, for a given shape of the profit curve, design the portfolio.

   (a) Construct a portfolio using only Calls where, at time $t = T$, the profit line is horizontal until price $\$60$. At that point, it has slope 3 until price $\$70$. Then, the line has slope 2 until price $\$100$. Next, it has slope zero until price $\$110$. It then has a slope of $-1$ until $\$120$. After that, it has slope zero.

(b) Construct a portfolio with the above behavior but using only Puts.

(c) Construct a portfolio with the above behavior using both Puts and Calls.

7. All of the following problems involve interest that is being compounded continuously.

(a) Compute the interest rate needed to double an investment every seven years.

(b) With a 10% interest, how long will it take to triple an initial investment.

(c) Irvin forgot what is the interest rate at his bank. All he knows is that after five years, his initial investment doubled. When will it triple? (Hint: There are two unknowns; one is the interest rate. Whenever there are two unknowns, two equations are needed. So, use the information about money at different times in an appropriate manner.)

(d) A person has money withdrawn from his savings account to be placed in his checking account at a continuous fixed basis at a rate of $1,000 a year. This person started with $20,000 placed in the savings account that has a 5% interest. How much will be in the savings account in one year? (Hint: A different expression for the growth of money needs to be derived. So, start with $\Delta M$ and, instead of the Equation 2.8 expression, determine what $\Delta M$ equals in this setting. Then, solve the equation.)

8. Two days before expiration date, Patty wants to sell a Call with strike price $100; i.e. she wants to go short on $C_{100,t}$. The interest rate is $r = 10\%$, and the current value of the stock is $120. Use the Put–Call Parity Equation to find a lower bound on the value of $C_{100,t}$.

9. Igor read in the press this morning that, for an expiration date of a year from now (with 5% interest) that $C_{60}(70, t) = 9$ and $P_{60}(70, t) = 4$. How can he use this information to make some money?

10. For this problem, assume that at this moment, $S = \$99, r = 0.02, C_{100}(99, T - \frac{1}{2}) - P_{100}(99, T - \frac{1}{2}) = \$8$, and $T = t + \frac{1}{2}$ year.

(a) Are there any arbitrage opportunities? If so, explain how to tackle this opportunity and determine how much money can be made.

(b) All sorts of events can happen in the next six months. Nevertheless, you are safe at expiration date. List the possibilities to explain how the profit you made remains made, and you can cover all debts.

11. For $S = 105, r = 0.05$, compute the following:

(a) Find the value of $P_{100}(105, T - 1) - C_{100}(105, T - 1)$ if $T - t = 1$ year.

(b) Find the value of $P_{105}(105, T - 1) - C_{105}(105, T - 1)$ if $T - t = 1$ year.

(c) Find the value of $P_{115}(105, T - 1) - C_{115}(105, T - 1)$ if $T - t = 1$ year.

(d) Find the value of $P_{100}(105, T - 2) - C_{100}(105, T - 2)$ if $T - t = 2$ years.

(e) Find the value of $P_{100}(105, T - 1) - C_{100}(105, T - 1)$ if $T - t = 1$ year and $r = 0.10$.

12. By using Equation 2.11, differences in the value of $C_E(S, t) - P_E(S, t)$ are computed.

(a) As shown above, if $S > E$, then the $C_E(S, t) - P_E(S, t)$ difference decreases with an increase in $E$. Why? Is it because $C_E(S, t)$ decreases in value? Or does $P_E(S, t)$ increase in value? Or both? Support your answer.

(b) The above analysis was done for $S$ greater than the present value of $E$. Carry out the same analysis if $S$ is less than the present value of $E$. That is, find what happens should

    i. $E$ increase (i.e., buying options with higher strike prices),
    ii. should $E$ decrease,
    iii. should $r$ increase,
    iv. should $T - t$ be larger, and
    v. should $S$ decrease.

# Chapter 3
# Modeling

## 3.1 Assumptions and Modeling

The football example demonstrated the wisdom of treating tacit assumptions with care: They may be false. Yes, the sum of the probabilities over all events equals one. But when dealing with different individuals, a selective sum of their "implicit probabilities" could differ from unity—which can either hurt or help you.

Similar concerns arise when modeling how asset prices change. The challenge is that the forces causing price changes probably are too complicated to be fully established. The best that can be expected is to identify situations where behavior is partly understood, which then is modeled with *assumptions*. Should the assumptions come close to being satisfied, conclusions may be reliable; if not, well, worry whether to trust the model.

Here is a practical concern: If a mathematical model could be created that captures all relevant aspects in pricing options, it most surely would be too complicated to analyze. Consequently, as typically true for modeling "real world" events, tradeoffs are made between *reality* and *simplifying assumptions*. The tradeoff? With overly simplified assumptions, the resulting models may be easy to use and analyze while contributing nothing of value about reality. On the other hand, models that adhere to reality may be impossible to solve and make predictions. Thus, adopted assumptions impose compromises in the predictive value of what is going on.

Assumptions must be taken seriously; it must be known what they do, and do not, mean; when they are, or are not, reasonably valid. By carefully identifying and understanding the assumptions, it becomes possible to appreciate which emerging predictions can be trusted.

© Springer Nature Switzerland AG 2019
D. G. Saari, *Mathematics of Finance*, Undergraduate Texts in Mathematics,
https://doi.org/10.1007/978-3-030-25443-8_3

## 3.1.1  Taylor Series

A way to underscore this point is with an analogy. Suppose you are on a job where you need to understand the behavior of a complicated function $f(x)$ near the value $x = a$. For an ugly example, suppose the goal is to understand the behavior of $y = Arctan\left[\ln(e^{\cos(x)} + \cot^2(x))\right]$ near $x = \frac{\pi}{2}$. No way!

Wouldn't it be nice if, for practical purposes, the complicated function could be replaced with a simple, approximating polynomial? This leads to the powerful tool of Taylor series; it replaces complicated functions with a simpler, easier to analyze polynomial approximations.

To review the approach for a single variable,[1] the goal is to find an approximating polynomial centered at $x = a$ that can replace the *actual* function $f(x)$. This requires finding the values of the coefficients $b_0, b_1, b_2, \ldots$ so that the approximation

$$f(x) \approx b_0 + b_1(x - a) + b_2(x - a)^2 + b_3(x - a)^3 + \ldots \tag{3.1}$$

is reasonably accurate near $x = a$. Rather than the full infinite series, in practice, as few terms as needed are used to create a polynomial that approximates $f(x)$ to a desired level.

To motivate the choice of coefficients, notice that when $x$ is at $a$, the left-hand side of Equation 3.1 is $f(a)$, while the right-hand side is $b_0$. So, a convenient way to kill off all of those polynomial terms is to set $x = a$: This reduces Equation 3.1 to $f(a) = b_0$, which suggests that the correct choice is

$$b_0 = f(a).$$

Similarly, a way to find the $b_1$ value is to discover how to eliminate the $(x - a)$ term of Equation 3.1. A natural approach is to differentiate both sides of Equation 3.1 to obtain

$$f'(x) \approx b_1 + 2b_2(x - a) + 3b_3(x - a)^2 + \ldots. \tag{3.2}$$

Using the same strategy to eliminate polynomial terms, set $x = a$ to obtain

$$b_1 = f'(a).$$

All other coefficients are found in the same manner. In particular, to find $b_j$, differentiate Equation 3.1 $j$ times (where $f^{(j)}(x)$ represents the $j^{th}$ derivative) to obtain

---

[1] While readers may know how to compute Taylor series for $y = f(x)$, what about for two variables such as $z = f(x, y)$ or three variables as with $u = f(x, y, z)$? Answers for multivariable settings can be obtained in the manner described next.

$$f^{(j)}(x) \approx (j!)b_j + (j+1)\ldots(2)b_{j+1}(x-a) + \ldots.$$

To drop all terms on the right-hand side, other than $b_j$, select $x = a$ to eliminate all $(x-a)^k$ terms. Solving both sides of the resulting equation for the remaining terms suggests that

$$b_j = \frac{f^{(j)}(a)}{j!}.$$

In particular, the Taylor series allows a smooth function $f(x)$ to be approximated by the $n^{th}$ degree polynomial

$$f(x) \approx f(a) + \sum_{j=1}^{n} \frac{f^{(j)}(a)}{j!}(x-a)^j, \tag{3.3}$$

where $f^{(j)}(a)$ is the $j$th derivative of $f$ evaluated at $x = a$.

The point is that an Equation 3.3 approximation is valid relative to basic assumptions: It is useful only for values of $x$ where $|x-a|$ is sufficiently small as determined by the error estimates (see any calculus book) and the desired degree of accuracy. To dramatically underscore this point, consider the bounded function $f(x) = \cos(x)$ with $a = \pi$ and $n = 2$. While the quadratic equation on the right-hand side of

$$\cos(x) \approx -1 + \frac{1}{2}(x-\pi)^2 \tag{3.4}$$

nicely approximates $f(x) = \cos(x)$ for $x$ values near $\pi$, the approximation totally collapses for larger $x$ values! After all, the quadratic approximation flies off to infinity as $x \to \infty$, while the value of $\cos(x)$ remains bounded.

**An Example**

To review the computations, let $f(x) = \ln(x)$ and $a = 1$. A Taylor series expansion requires computing the derivatives of $f(x)$, which are

$$f(x) = \ln(x), \ f'(x) = \frac{1}{x}, \ f''(x) = -\frac{1}{x^2}, \ldots, \ f^{(n)}(x) = (-1)^{n+1}\frac{(n-1)!}{x^n}.$$

These derivatives are evaluated at the specified $x = 1$, which is easy because the $x^j$ terms in the denominators will equal unity. As such,

$$f(1) = \ln(1) = 0, \ f'(1) = 1, \ f''(1) = -1, \ldots, \ f^{(n)}(1) = (-1)^{n+1}(n-1)!$$

The $(-1)^{n+1}$ term captures the alternating change of sign of the derivative where $f'$ is positive, $f''$ is negative, and so forth. (Thus, if $n$ is odd, then $f^{(n)}$ is positive as required by the $(-1)^{n+1}$ multiplier.)

Because

$$b_j = \frac{f^{(j)}(1)}{n!} = \frac{(-1)^{n+1}(n-1)!}{n!} = (-1)^{n+1}\frac{1}{n},$$

the Taylor series representation is

$$\ln(x) \approx \sum_{n=1}^{\infty} \frac{(-1)^{n+1}}{n}(x-1)^n. \tag{3.5}$$

To illustrate by computing $\ln(1.1)$, let $x = 1.1$ in Equation 3.5 to obtain

$$\ln(1.1) = \sum_{n=1}^{\infty} \frac{(-1)^{n+1}}{n}(1.1-1)^n.$$

As the $(1.1-1)^j = \frac{1}{10^j}$, an answer accurate to five decimal places would have $n = 5$, or

$$\ln(1.1) \approx .1 - \frac{0.01}{2} + \frac{0.001}{3} - \frac{0.0001}{4} + \frac{0.00001}{5} - \frac{0.000001}{6}.$$

## Another Example

To compute the value of $e^{-0.02}$ without a calculator, observe that $a = 0$ is close to $x = -0.02$. This suggests finding the Taylor series approximation for $f(x) = e^x$ about $a = 0$.

The computations are trivial; any derivative of $e^x$ is $e^x$; i.e., $f^{(n)}(x) = e^x$. Moreover, $e^a = e^0 = 1$, so $f^{(n)}(0) = 1$ for all $n$. Therefore the Taylor series is

$$f(x) = e^x \approx 1 + 1(x-0) + \frac{1}{2!}(x-0)^2 + \frac{1}{3!}(x-0)^3 + \frac{1}{4!}(x-0)^4 + \cdots$$
$$= \sum_{j=0}^{\infty} \frac{x^j}{j!}. \tag{3.6}$$

To approximate the $e^{-0.02}$ value, let $x = -0.02$ in Equation 3.6 to obtain

$$e^{-0.02} \approx 1 - 0.02 + \frac{1}{2!}(-0.02)^2 + \frac{1}{3!}(-0.02)^3 + \cdots$$
$$= 1 - 0.02 + 0.0002 - 0.0000013.. + \cdots,$$

which is close to (but slightly larger than) 0.98.

**WGAD**

As for the "Who gives a darn?" concern, the reader should. Imagine Helena, needing to quickly reply to a client's question, must compute the present value of $100 for a year from now at 2% interest compounded continuously. Should Helena recall from Equation 3.6 that $e^x \approx 1 + x$ for small $x$ values, she could respond with the reasonable approximation of

$$100e^{-0.02} \approx 100[1 - 0.02] = 100[0.98] = 98.$$

Similarly, the present value of $50 with 6% interest over a half year is $50e^{-0.06(0.5)} = 50e^{-0.03} \approx 50(1 - 0.03) = \$48.50$. With large values such as $x = 1$, however, the $e^x \approx 1+x$ approximation loses validity. After all, $e^1 \approx 2.71828$ is larger than $1 + 1 = 2$. This only means that more Equation 3.6 terms are required to have an appropriate approximation.

### 3.1.2  More Variables

To analyze the behavior of $f(x, y) = e^{\cos(x)}/y$ near $x = \frac{\pi}{2}$ and $y = 2$, try a simpler polynomial approximation. It would be

$$f(x, y) \approx b_{0,0} + b_{1,0}(x - a_1) + b_{0,1}(y - a_2) +$$
$$+ b_{2,0}(x - a_1)^2 + b_{1,1}(x - a_1)(y - a_2) +$$
$$+ b_{0,2}(y - a_2)^2 + \cdots + b_{j,k}(x - a_1)^j (y - a_2)^k + \dots,$$

where $a_1 = \frac{\pi}{2}$ and $a_2 = 2$.

Find the $b_{j,k}$ coefficients as above. Namely, differentiate $f(x, y)$ and the series the appropriate number of times to eliminate the variables in the $b_{j,k}(x - a_1)^j (y - a_2)^k$. Doing so leaves $\frac{\partial^{j+k} f(x,y)}{\partial x^j \partial y^k} \approx j!k!b_{j,k} +$ polynomial terms with $(x - a_1)^r (y - a_2)^s$. Eliminate these extra polynomial terms by setting $x = a_1$, $y = a_2$, which leads to

$$b_{j,k} = \frac{1}{j!k!} \left( \frac{\partial^{j+k} f(a_1, a_2)}{\partial x^j \partial y^k} \right). \tag{3.7}$$

Illustrating with $f(x, y) = e^{\cos(x)}/y$, because a second order approximation is

$$f(x, y) \approx f(a_1, a_2) + \frac{\partial f(a_1,a_2)}{\partial x}(x - a_1) + \frac{\partial f(a_1,a_2)}{\partial y}(y - a_2) + \frac{1}{2}\frac{\partial^2 f(a_1,a_2)}{\partial x^2}(x - a_1)^2$$
$$+ \frac{\partial^2 f(a_1,a_2)}{\partial x \partial y}(x - a_1)(y - a_2) + \frac{1}{2}\frac{\partial^2 f(a_1,a_2)}{\partial y^2}(y - a_2)^2, \tag{3.8}$$

the second order approximation for the above $f(x, y)$ is

$$\frac{e^{\cos(x)}}{y} \approx 1 - \frac{1}{2}\sin(\frac{\pi}{2})e^{\cos(\frac{\pi}{2})}(x - \frac{\pi}{2}) - \frac{1}{2^2}e^{\cos(\frac{\pi}{2})}(y - 2)$$
$$+ \frac{1}{2}[-\cos(\frac{\pi}{2}) + \sin^2(\frac{\pi}{2})]\frac{1}{2}e^{\cos(\frac{\pi}{2})}(x - \frac{\pi}{2})^2 - \frac{1}{2^2}\sin(\frac{\pi}{2})e^{\cos(\frac{\pi}{2})}(x - \frac{\pi}{2})(y - 2)$$
$$+ \frac{1}{2}[\frac{2}{2^3}e^{\cos(\frac{\pi}{2})}](y - 2)^2$$

which reduces to a more civilized

$$\frac{e^{\cos(x)}}{y} \approx 1 - \frac{1}{2}(x - \frac{\pi}{2}) - \frac{1}{4}(y - 2) + \frac{1}{4}(x - \frac{\pi}{2})^2 - \frac{1}{4}(x - \frac{\pi}{2})(y - 2) + \frac{1}{8}(y - 2)^2.$$

Remember, this approximation is valid only near $x = \frac{\pi}{2}$, $y = 2$.

### 3.1.3  Back to Modeling Approximations

Understanding whether a Taylor series approximation reasonably captures the behavior of a function extends to understanding whether simplifying assumptions represent how an asset's prices change. Stated more strongly, a model's predictive value depends on whether the model's underlying assumptions reasonably reflect market behavior. Settings where assumptions fail to do so identify topics for research and/or possible arbitrage advantage.

The message: Treat models and their constraints as

   *simplifying assumptions and approximations that can be wrong.*

Just as with Taylor series, there is a need to understand where models fail, and where they are reasonably accurate.

## 3.2  Efficient Market Hypothesis

What affects market prices? It is worth taking a break to consider relevant factors. Jennifer, who has managerial interests, might accurately argue that the choice of a company's executives, corporate governing structures, election outcomes, competitor strategies, and current prices are important, while Samuel, with an agricultural background, might stress the relevance of weather conditions such as droughts, hurricanes, and blizzards, along with tariffs and possible trade wars. Who knows, maybe even what songs are currently popular might have an impact.

Here is the problem: How can *any* model be developed to incorporate all of these conditions? Even if a model could be created, it most surely would be beyond any realistic ability to analyze.

Simplifying assumptions are required: One choice is called the *Efficient Market Hypothesis* (EMH). A basic premise of a simple version of the EMH is that, with so many smart, rapidly reacting people on the market, certain types of market information already have been sucked out and used to determine current prices. Namely,

1. All information about prices in the past already is reflected in the current price of the asset.
2. All new information about the market is immediately incorporated into the price.

A real purpose of the EMH is to eliminate complicating concerns when creating models: The choice of the company's executives? Don't worry as it already is built into the price. Climate change? Not a concern because the price already captures this feature. What remains is the price of the asset.

From a modeling perspective, assumption 1 is delightful; it allows complications of the past history to be blissfully ignored. But while it simplifies the analysis, does the assumption always make sense? For instance, suppose the news reports that Apple stock is down so many points. If the EMH holds, then this stock would be treated the same as another one that dropped this morning. But Apple has a reputation of being a 800 pound gorilla, so rather than accepting item 1, perhaps this news signals an opportunity to buy or that there are worries for the full market. Similarly Apple's demonstrated ability to overcome all sorts of adversity surely plays a role in whether we should believe the stock price will increase. But the EMH states that the current price already reflects all of this information.

Similarly, the helpful second assumption permits ignoring what *kind of information* is on the market. Is it valid? Suppose, for instance, that Bernie spent last summer enjoying the delights of Finland where he discovered that the Finns need inexpensive, light weight but *warm* clothing. Suppose he also read in today's *Textile Magazine* that Professor X discovered how to convert used blue books from math of finance exams into warm clothing. There is a tantalizing connection, but it is unlikely that this public information has affected today's price. Why? Most people are unaware of the potential market for the new discovery perhaps coupled with a deep aversion to exams and exam books: They have not made a connection between the two events.

A purpose of these arguments is to encourage the reader to seriously think about where and when the EMH is realistic and useful. Does this approximating assumption, like Taylor series, apply only to "short run" settings where everything is fairly calm?

The usefulness of the EMH depends on whether *you* represent that hidden premise where "there are so many smart, rapidly reacting people on the market." The difference is whether you are a leader on the forefront—one of those smart, rapidly reacting people—or a follower; the former probably can ignore the EMH, while the latter should embrace it.[2] Beyond expecting aspects of the EMH to hold during

---

[2]At Northwestern University, I would invite Arthur Pancoe, a highly successful investor, to visit my class. (A measure of his success is a headline in a 1988 issue of Money Magazine, "Take Two of Arthur Pancoe's Drug Stocks and You May Be Rich in the Morning.") He would ask the students to guess what were his daily readings. To their surprise, the readings included non-finance outlets such as *Science* and *Nature*, which gave him insights into what was being discovered. He also

periods of calm, the reader is encouraged to explore whether there are other settings. Doing so with EMH and other assumptions can provide a personal advantage.

As a summary, treat EMH as a convenient assumption—needed for the construction of a tractable mathematical model. Is it valid in general? Of course not! Just as Equation 3.4 approximation for $\cos(x)$ is useful for certain settings, it can lead you astray more generally. Similarly, the EMH is a reasonable expectation for certain surroundings—but not for others.

Of "open disclosure" importance, the EMH is a basic hypothesis for what follows. The value of assumptions is that they allow insights and answers for appropriate settings. It is to the reader's advantage to determine when and whether this is true for what follows.

### 3.2.1  Modeling

If $S$ is the price of asset $S$ (yes, same notation for both), the goal is to predict price changes $\Delta S$. Relative price changes are important, so a first step is to determine what causes changes in $\Delta S/S$. The EMH simplifies that analysis by emphasizing the current price.

For insight of what to do, recall how money in a bank changes according to the interest. If $M$ is the money in the bank, $r$ is the interest rate, and $\Delta t$ is a small increment of time, the accompanying change in money is

$$\Delta M = rM\Delta t. \tag{3.9}$$

What makes Equation 3.9 format such a commonly used expression is that the right-hand side often represents the first term in a Taylor series approximation of an unknown force. Change in temperature in a room of $T^*$ degrees is given by $\Delta T = F(T)\Delta t$. Fine, but what if the function $F$ is not known? One approach is to replace the unknown $F(T)$ with what would be the first term of its Taylor series or $F(T) \approx \mu(T - T^*)$. This leads to Newton's law of cooling

$$\Delta T = \mu(T - T^*)\Delta t. \tag{3.10}$$

The unknown constant $\mu$ cannot come from the modeling, so it is determined from data.

Such a linear model asserts that the rate of change in an object is proportional to how much is currently available. For instance, for a *first approximation*, a change in

---

declared that he did not accept the EMH. Of course not: Pancoe was a leader, not a follower, so the basic assumption behind the EMH did not apply to him. He was one of those reacting people behind the premise.

population, $\Delta P$, is proportional (given by the constant $\alpha$) to the current population ($P$) multiplied by the change in time.[3] That is,

$$\Delta P = \alpha P \Delta t,$$

or

$$\frac{\Delta P}{P} = \alpha \Delta t. \tag{3.11}$$

This standard argument suggests that for a *first approximation* of how prices of an asset change, it is reasonable to assume, at least for a short time span, that it is proportional to the current price. For a start, assume that

$$\frac{\Delta S}{S} = \mu \Delta t. \tag{3.12}$$

The proportionality constant $\mu$ measures a deterministic effect, where, as with Newton's law of cooling (Equation 3.10), its value is derived from available data about the asset. A particularly simple assumption is that $\mu$ is a constant; more accurate models require $\mu(S, t)$ to reflect how the price affects the "drift," etc.

Here is a problem: Changes in an asset's value are *not* strictly deterministic; the adjusting forces most surely include random effects needed to capture what cannot be accurately modeled. This requires Equation 3.12 to include random, unanticipated motion. To prepare for this discussion, the reader is encouraged to review facts from Chapter 1, which are needed for the discussion about the *normal distribution.*

## 3.2.2   Random Variables

Recall that the *expected value* of a random variable $X$ is

$$E(X) = \sum_{j=1}^{n} p_j X(j) \text{ or } E(X) = \int_{-\infty}^{\infty} X(x) f(x) dx,$$

---

[3]Thomas Malthus (1786–1834) used an Equation 3.11 type expression to analyze population growth. Similar to Equation 3.4, this approximation is reasonable "in the small" but definitely not in general. Indeed, this incorrect representation led to the several century Malthusian debate. One correction was to replace the linear term with a quadratic expression: Use not just the first, but the first two terms of a Taylor series expansion of $aP + bP^2 = P(a + bP)$. Doing so created a search for explanations of the coefficients, such as carrying capacity, death rates, and the logistic equation. The message: Be careful; compare predictions with data.

with the variance given by

$$\mathrm{Var}(X) = E([X - E(X)]^2) = E(X^2 - 2XE(X) + (E(X))^2)$$
$$= E(X^2) - [E(X)]^2, \tag{3.13}$$

where $\sigma = \sqrt{\mathrm{Var}(X)}$, the standard deviation, serves as a unit of "distance."

To review the terms with a penny spinning on its edge, where the probability of getting a H is 0.3, if $X(H) = 1$, $X(T) = 0$, then $X$ counts Heads:

$$E(X) = 0.3X(H) + 0.7X(T) = 0.3,$$

and

$$\mathrm{Var}(X) = E(X^2) - [E(X)]^2 = [0.3X^2(H) + 0.7X^2(T)] - 0.09 = 0.3 - 0.09.$$

Also recall that the standard form is

$$Z = \frac{X - \mu}{\sqrt{\mathrm{Var}(X)}} = \frac{X - \mu}{\sigma},$$

which, with the above numbers, is

$$Z = \frac{X - \mu}{\sigma} = \frac{X - 0.3}{\sqrt{0.21}}.$$

**Normal Distribution**

For another example, let $U \sim N(0, 1)$, where $N(0, 1)$ represents the PDF of the normal distribution with variance 1. The PDF is

$$f(x) = \frac{1}{\sqrt{2\pi}} e^{-\frac{x^2}{2}};$$

the constant $\frac{1}{\sqrt{2\pi}}$ means that the area under the curve $y = e^{-\frac{x^2}{2}}$ is $\sqrt{2\pi}$. Thus $U$'s standard form is $Z = \frac{U-0}{1} = U$; i.e., random variable $U$ already is in its standard form. The importance of this comment is that when comparing $U$ with another random variable in standard form, what is being compared are two random variables in standard form.

To compute the probability that $U \leq t$ where $t$ is some number is easy

$$P(U \leq t) = \int_{-\infty}^{t} f(x)\, dx = \frac{1}{\sqrt{2\pi}} \int_{-\infty}^{t} e^{-\frac{x^2}{2}}\, dx. \tag{3.14}$$

The value of Equation 3.14 integral is given in "standard normal distribution tables," which are readily available everywhere; e.g., the internet: They specify the $Z$ value (the value of this integral) for given $t$ (standard deviation) terms.


### 3.2.3  Back to Finance

The puzzle to be solved is to find an appropriate random variable to model the random effects on the prices. This random term is intended to reflect the accumulated, or aggregated, effects of many individuals, so insight may be gained by reviewing sums of random variables, such as

$$Y_n = \sum_{j=1}^{n} X_j.$$

Illustrating with the spinning coin example, if it is spun on edge 100 times, and $X_j$ is the random variable indicating whether the outcome is H on the $j^{th}$ spin, then $Y_{100} = \sum_{j=1}^{100} X_j$ is the number of Heads.

The analysis of $Y_n = \sum_{j=1}^{n} X_j$, a sum of random variables, is carried out as above: Find the standard form of $Y_n$, which is

$$Z_n = \frac{Y_n - E(Y_n)}{\sqrt{\mathrm{Var}(Y_n)}} = \frac{\sum_{j=1}^{n} X_j - E(\sum_{j=1}^{n} X_j)}{\sqrt{\mathrm{Var}(\sum_{j=1}^{n} X_j)}}. \tag{3.15}$$

If, for instance, an example has $Z_n \leq -0.4$, then the summation is below the expected value by at least 0.4 standard deviations.

The probabilities of various events can be computed by using the amazing *central limit theorem*. First the assumptions (which the reader should consider with care): The theorem requires the $X_j$ random variables to be iid, which means "independent and identically distributed."

1. The "identically distributed" requires each $X_j$ to have the same PDF, so each $X_j$ has the same mean $\mu$ and standard deviation $\sigma$. Flipping the same penny many times is an example.
2. The "independence condition" means that the outcome of each $X_j$ is independent of what happens with any other random variable. A classic example is tossing a die $n$ times: Each throw has no influence on what happens with any other throw. For our purposes, the independence assumption has the powerful consequence that

$$\mathrm{Var}(\sum X_j) = \sum \mathrm{Var}(X_j). \tag{3.16}$$

As such, the iid assumption about these terms simplifies Equation 3.15 to

$$Z_n = \frac{\sum_{j=1}^{n} X_j - E(\sum_{j=1}^{n} X_j)}{\sqrt{\text{Var}(\sum_{j=1}^{n} X_j)}} = \frac{\sum_{j=1}^{n} X_j - n\mu}{\sqrt{n\sigma^2}} = \frac{\sum_{j=1}^{n} X_j - n\mu}{\sqrt{n}\sigma}.$$

(3.17)

If $\mu$ and $\sigma > 0$ have finite values, the central limit theorem surprisingly asserts that, as $n$ assumes larger values, the likelihood that $Z_n \leq t$ (the likelihood that the data is bounded above by $t$ standard deviations) is the same as the likelihood that $U \leq t$ where $U \sim N(0, 1)$ (i.e., $U$ is the above normal distribution that already is in standard form). Amazing! Moreover, this assertion holds for *any choice of PDFs* for the $X_j$ random variables!!

Solving problems becomes immediate: Convert a specified $\sum_{j=1}^{N} X_j \leq a$ issue into its standard form. For an illustration, return to where the summation $\sum_j X_j$ counts the number of $H$s that arise when spinning the coin on its edge. (The PDF is $p(H) = 0.3$, $p(T) = 0.7$.) If the coin is spun 100 times, what is the likelihood of obtaining no more than 20 Heads? That is, what is the likelihood that $Y_{100} = \sum_{j=1}^{100} X_i \leq 20$?

Each spin is iid, and $\mu = E(X_i) = X(H)p(H) + X(T)p(T) = 0.3$, where, as computed above, $\text{Var}(X_i) = E(X_i^2) - (E(X_i))^2 = 0.3 - 0.09 = 0.21$. Next, use this information to convert the $Y_{100} \leq 20$ concern into a $Z_{100} \leq t$ format, and then lean on the central limit theorem for an answer.

The needed terms to convert $Y_{100} \leq 20$ to $Z_{100} \leq t$ are specified in Equation 3.17. That is, first subtract $E(Y_{100})$ from both sides of $Y_{100} \leq 20$ inequality and then divide both sides by $\sqrt{\text{Var}(Y_{100})}$ to obtain

$$P(\sum_{j=1}^{100} X_i \leq 20) = P(\sum_{j=1}^{100} X_i - E(\sum_{j=1}^{100} X_j) \leq 20 - 30)$$
$$= P(\frac{\sum_{j=1}^{100} X_i - 100\mu}{\sqrt{\text{Var}(Y_{100})}} \leq \frac{20-30}{4.58})$$
(3.18)
$$= P(Z_{100} \leq -2.18).$$

The central limit theorem asserts that with large values of $n$, the probability for Equation 3.18 approaches the value

$$P(U \leq -2.18) = \frac{1}{\sqrt{2\pi}} \int_{-\infty}^{-2.18} e^{-\frac{x^2}{2}} \, dx = 0.0146.$$

The small probability value of only 0.0146 comes from normal distribution tables by checking the value for the standard deviation of $-2.18$.[4]

**Theorem 1 (Central Limit Theorem)** *Suppose that* $X_1, \ldots, X_n, \ldots$ *is a sequence of independent, identically distributed random variables with finite* $E(X_j) = \mu$ *and* $\text{Var}(X_j) = \sigma^2 > 0$. *Let* $Z_n = \frac{\sum_{j=1}^{n} X_j - E(\sum_{j=1}^{n} X_j)}{\sqrt{\text{Var}(\sum_{j=1}^{n} X_j)}}$. *Then*

---

[4] Rather than $Y_n = \sum X_i$, the average $Y_n^* = \frac{1}{n} \sum_{j=1}^{n} X_j$ is often used. No problem; do the same by finding the associated $Z$ representation, which leads to the same conclusion.

$$\lim_{n \to \infty} P(Z_n \le t) = \frac{1}{\sqrt{2\pi}} \int_{-\infty}^{t} e^{\frac{-x^2}{2}} \, dx, \quad -\infty < t < \infty. \tag{3.19}$$

As a parting example, suppose the average starting salaries of graduates from Upper University is \$40,000 with a standard deviation of $\sigma = \$1,200$. Suppose that Enrique, the university's PR person, believes the average salary is higher. As such, he took an arbitrary sample of 100 graduates, and found that their average starting salary is over \$42,000. As this value could be just luck, Enrique wants to find the likelihood that, if the stated information is correct, the average starting salary of a randomly drawn sample of 100 students is at least \$42,000.

The problem is to find $P(Y_{100} = \frac{1}{100}\sum_{j=1}^{100} X_j > 42,000)$ where $X_j$ is the $j^{th}$ person's starting salary. Re-expressing this problem in terms of $Z_{100}$, we have $Z_{100} = \frac{Y_{100}-40,000}{120} > \frac{42,000-40,000}{120} = 1\frac{2}{3}$.

To explain the denominator of 120, the standard deviation, recall that $\text{Var}(Y_{100}) = \text{Var}(\frac{1}{100}\sum_{j=1}^{100} X_j) = \frac{1}{100^2}\text{Var}(\sum_{j=1}^{100} X_j)$. The fact the $X_j$ are independent means that the variance of the sum equals the sum of the variances (which is partly explained at the beginning of Chapter 4). Thus,

$$\text{Var}(Y_{100}) = \frac{1}{100^2} \sum_{j=1}^{100} \text{Var}(X_j) = \frac{1}{100}\text{Var}(X_j).$$

The standard deviation, or square root, is $\frac{1}{10}\sqrt{\text{Var}(X_j)} = \frac{1}{10}\sigma = 120$.

After all of these computations, it follows from the central limit theorem that an approximate value of the probability is

$$P(Y_{100} > 42,000) = P(\frac{Y_{100}-40,000}{120} > \frac{10}{6}) \approx \frac{1}{\sqrt{2\pi}} \int_{\frac{10}{6}}^{\infty} e^{-\frac{x^2}{2}} \, dx = 0.0479.$$

So, there is about a 1 in 20 chance that Enrique's survey came from the university's specified distribution.

## 3.2.4 Random Effects

As this example concerning graduates indicates, the central limit theorem is incredibly forgiving: After all, there is no information whatsoever about the probability distribution of starting salaries for the UU graduates.

Here is a surprise: the choice of the random variable, whether its PDF is known or not, does not matter! This permits the theorem to handle an assortment of concerns. In particular, the random terms affecting changes of prices emerging from the behavior of a large number of people, where the $j^{th}$ person is captured by $X_j$, often are modeled in this manner.

More specifically, the random change in the asset price is modeled with

$$\sigma \Delta X,$$

where $\sigma$ is called the *volatility*. The $\Delta X$ term is assumed to be $\Delta X \sim N(0, \Delta t)$. This means that $\Delta X$ is

- a random variable with a normal distribution
- with mean zero
- and variance $\Delta t$.

It is reasonable to believe that the variance of a random change in the asset price over the $\Delta t$ time interval is a multiple of $\Delta t$ (after all, this is the length of time over which something can happen), where the multiple reflects market activity. According to the $\sigma \Delta X$ choice, it is $\sigma^2 \Delta t$, which captures the market's volatility.

With these assumptions, a price change equation becomes

$$\frac{\Delta S}{S} = \mu \Delta t + \sigma \Delta X, \tag{3.20}$$

or

$$\Delta S = \mu S \Delta t + \sigma S \Delta X. \tag{3.21}$$

Keeping with the counsel of remaining vigilant when modeling, what does the above mean? Is this assumption, as applied to finance, reasonable? Do you *really* *believe* that what one person does is independent of another? As we move along, doubt will be cast on the assumptions validity.

## 3.3  Interpretation

It is tempting to appeal to calculus to solve Equation 3.20. Using the definition of an integral, perhaps a quick way to solve for $S(t)$ is to find the best possible approximation of

$$\sum \frac{\Delta S}{S} = \mu \sum \Delta t + \sigma \sum \Delta X. \tag{3.22}$$

The best possible value from expressions of this type, as $\Delta t \to 0$, is

$$\int \frac{1}{S} dS = \mu \int dt + \sigma \int dX. \tag{3.23}$$

There is a problem; the second integral on right-hand side needs to be interpreted. To explain, $\int_a^b f(x)\, dx$ is the best possible value obtained by creating sums of rectangles with a small base $\Delta x$. Doing this with $\int dX$, for a small $\Delta t$ value (the variance), requires computing the corresponding $\Delta X$ value. But $\Delta X$ is a *random variable* with a normal distribution, so the value could be positive, negative, or even zero, and it most surely will change when recomputed. Thus, there is no reason to expect that these summations tend toward some value. The best that can be expected from this stochastic integral is for the answer to be a probability distribution.

### 3.3.1 Probability Distribution

So far we cannot solve Equation 3.21, but at least we can find some properties. The expected value of $\Delta S$ is

$$E(\Delta S) = \mu S \Delta t + \sigma S E(\Delta X) = \mu S \Delta t, \tag{3.24}$$

which holds because $\mu S \Delta t$ is a deterministic term (so it equals its expected value), and, by assumption, $E(\Delta X) = 0$. According to Equation 3.24, the expected incremental price change is determined by the drift term.

This fact that $E(\Delta X) = 0$ simplifies the computation of the variance of $\Delta S$. It is

$$
\begin{aligned}
\text{Var}(\Delta S) &= E((\Delta S)^2) - (E(\Delta S))^2 \\
&= E(\sigma^2 S^2 (\Delta X)^2 + 2\sigma \mu S^2 \Delta X \Delta t + \mu^2 S^2 (\Delta t)^2) - \mu^2 S^2 (\Delta t)^2 \\
&= E(\sigma^2 S^2 (\Delta X)^2) + \Delta t 2\sigma \mu S^2 E(\Delta X) = \sigma^2 S^2 \Delta t.
\end{aligned}
\tag{3.25}
$$

The only above term not canceling is $E(\sigma^2 S^2 (\Delta X)^2) = \sigma^2 S^2 E((\Delta X)^2)$, where, because $E(\Delta X) = 0$, $E((\Delta X)^2) = \text{Var}(\Delta X) = \Delta t$.

The standard deviation of $\Delta S$ is $\sigma S \sqrt{\Delta t}$, which means that the volatility—the way in which the incremental price change swings around the drift—has a standard deviation of $\sigma S \sqrt{\Delta t}$. For sake of intuition, this volatility depends on the current price, S, the standard deviation of the random term, $\sqrt{\Delta t}$, and an extra term $\sigma$ that captures the volatility of the market.

While these arguments provide useful information, they fail to identify the PDFs that represent the changes in prices or value of an asset. This is computed in the next chapter.

## 3.4 Exercises

1. Find examples from current events where it is arguable that the Efficient Market Hypothesis is correct, and where it probably is false. (That is, check the business pages.)

2. Find the Taylor series approximation for $f(x) = \cos(x)$ about $x = \pi$. With this approximation, find the value of $\cos(\pi - 0.1)$ that is valid for six decimal points.
3. By use of Taylor series, find an approximate value for $e^{1.01}$ that is accurate to three decimal places.
4. A cube is built with inside dimensions of 10 inches. The material is 0.2 inches thick. Use a Taylor series approximation to find the approximate volume of material used.
5. *This problem is important for our development in the next chapter. Be sure you understand it!*

   Suppose you want to find an approximate representation for a function $z = f(x, y)$ that is valid near $x = 1$, $y = 2$. Namely, you want to have

   $$f(x, y) \approx b_0 + b_{1,0}(x - 1) + b_{0,1}(y - 2) +$$
   $$b_{2,0}(x - 1)^2 + b_{1,1}(x - 1)(y - 2) + b_{0,2}(y - 2)^2 + \ldots .$$

   a. Use the ideas from the beginning of this chapter to find the coefficients.
   b. What are the four coefficients for the third order terms that have the form $(x - 1)^3$, $(x - 1)^2(y - 2)$, $(x - 1)(y - 2)^2$, $(y - 2)^3$?
   c. How many coefficients are needed for the fourth order terms?
   d. Find the Taylor series approximation up to order two near $x = 0$, $y = \pi$ of $f(x, y) = e^x \cos(y)$.
   e. For a function $w = f(x, y, z)$, what is the Taylor series approximation up to quadratic terms around $x = a_1$, $y = a_2$, $z = a_3$.

6. Suppose you have a fair die. (Namely, when the die is tossed, it is equally likely that any of the six numbers—1, 2, ..., 6—will appear.) You are playing a game where, if a 1 or 2 appears, you win \$1. If a 3 or 4 appears, you win \$2. If a 5 or 6 appears, you lose \$3. If the game is played a hundred times, what is the likelihood of winning ten or more dollars?
7. Suppose the only game in town involves flipping a fair coin (so Heads and Tails are equally likely), with a \$x bet. If Heads comes up, the payoff is \$0.9x; if Tails comes up, you lose the \$x. You have \$10,000, and must win at least \$5,000 by tomorrow morning to pay off a debt to a mean dude.

   a. Compute the likelihood of winning at least \$5000 by making a single bet of \$10,000.
   b. Compute the likelihood of winning at least \$1000 by playing the game 10,0000 times and betting a dollar each time. What is the likelihood of not losing money?
      Message learned?

8. Suppose a PDF $f(x)$ is zero for $x < 0$, $C x$ for $0 \le x \le 1$, and zero for $x > 1$. Suppose a game is played where the random variable $X(x) = x$ specifies the outcome. If the game is played 100 times, find the likelihood that the sum of the numbers is between 0 and 50.

# Chapter 4
# Some Probability

## 4.1 Review of Some Probability

Imagine what personal advantage could accrue if we knew, weeks in advance, tomorrow's price of IBM or Uber stock! Such information would identify what Calls, Puts, or other financial activities to put into place to support an early retirement.

Finding such information means that, in some sense, Equation 3.21 must be solved. This objective requires understanding Itô's Lemma, which is introduced in this chapter. But first, certain basic notions from probability are reviewed. A way to introduce them is in terms of the pains of gambling.

Suppose the only game in town is to spin a Norwegian Kroner on edge where you must take Heads. For reasons described earlier, the probability of Heads surely differs from 0.50. For sake of analysis, assume that

$$p(H) = 0.4, \quad p(T) = 0.6$$

and the random variable is $X(H) = 1$, $X(T) = -1$. That is, if Heads appear, a dollar is won; if Tails appear, a dollar is lost.

The expected value per spin of this coin, or expected winning, is

$$E(X) = p(H)(1) + p(T)(-1) = 0.4 - 0.6 = -0.2.$$

with variance

$$\text{Var}(X) = E(X^2) - (E(X))^2 = p(H)(1)^2 + p(T)(-1)^2 - (0.2)^2 = 1 - 0.04 = 0.96.$$

Suppose you decide to gamble and enjoy a winning streak, but then you suffer a spell of bad luck. The almost universal temptation is to continue to play until your

© Springer Nature Switzerland AG 2019
D. G. Saari, *Mathematics of Finance*, Undergraduate Texts in Mathematics,
https://doi.org/10.1007/978-3-030-25443-8_4

luck changes in order to recover the lost money. All of this is personally justified, of course, with the standard promise of "Then I'll quit."

To examine the potential consequences of continuing to gamble, each play, $X_i$, is independent of all others, and the probability is identically distributed. Consequently, the assumptions needed to invoke the central limit theorem are satisfied.

If $Y_n = \frac{1}{n} \sum_{i=1}^{n} X_i$ where $n$ is the number of spins, then $Y_n$ is the average winning per spin. This "average" interpretation suggests that, in some manner, attained $Y_n$ values should be related to the expected value $E(X_i) = -0.2$. To see whether this is true, suppose $n = 100$, and compute the probability that

$$P(-0.30 \le Y_{100} \le -0.10). \tag{4.1}$$

Equation 4.1 determines the likelihood that, on average, our gambler is going broke.
To compute this value, according to the CLT,

$$\lim_{n \to \infty} P(a \le \frac{Y_n - E(Y_n)}{\sqrt{\text{Var}(Y_n)}} \le b) = \frac{1}{\sqrt{2\pi}} \int_a^b e^{-\frac{x^2}{2}} dx. \tag{4.2}$$

So compute the $E(Y_n)$ and $\text{Var}(Y_n)$ values and use them with Equation 4.1 to determine Equation 4.2 $a$ and $b$ values.

The first computation uses

$$E(Y_n) = E(\frac{1}{n} \sum_{j=1}^{n} X_j) = \frac{1}{n} \sum_{j=1}^{n} E(X_j).$$

Because each $X_j$ has the same distribution, all share the same mean and variance. For our problem, $E(X_j) = -0.2$, so

$$E(Y_n) = \frac{1}{n} \sum_{j=1}^{n} -0.2 = \frac{n}{n}(-0.2) = -0.2.$$

Consequently Equation 4.1 can be expressed as

$$P(-0.30 \le Y_{100} \le -0.10) = P(-0.30-(-0.2) \le Y_n - E(Y_n) \le -0.10-(-0.2)),$$

which defines the equivalent problem of finding

$$P(-0.10 \le Y_n - E(Y_n) \le 0.10).$$

To compute the variance, notice that if $Y = aX$ where $a$ is a scalar, then $E(Y) = aE(X)$. Of more interest,

$$\text{Var}(Y) = E(Y^2) - (E(Y))^2 = E(a^2 X^2) - (a E(X))^2 = a^2 \text{Var}(X).$$

Thus, when factoring a constant out of the variance, it is squared. This makes sense: $aX$ increases the $X$ outcomes by the multiple $a$, and the variance is "distance squared" of the data from the mean.

To appreciate potential difficulties of dealing with a sum of variables, compute the variance of $Z_1 + Z_2$ where $Z_1$ and $Z_2$ are the random variables with mean zero. Here,

$$\text{Var}(Z_1 + Z_2) = E((Z_1 + Z_2)^2) = E(Z_1^2) + 2E(Z_1 Z_2) + E(Z_2^2). \qquad (4.3)$$

The first and last terms are, respectively, the variance of $Z_1$ and $Z_2$. (This is because $E(Z_1) = E(Z_2) = 0$.)

The messy middle term can cause problems; this is where the assumption of *independence* is important. With independence, $E(Z_1 Z_2) = E(Z_1)E(Z_2)$. Again, this makes sense. If the PDFs for $E_1$ and $E_2$ are, respectively, $h(x)$ and $k(y)$, then independence means that $h$ values do not depend on $y$ and $k$ values are not influenced by $x$ values. Consequently, from calculus,

$$E(Z_1 Z_2) = \int \int h(x) k(y) \, dx dy = \left( \int h(x) \, dx \right) \left( \int k(y) \, dy \right) = E(Z_1)E(Z_2).$$

Returning to Equation 4.3, the fact $E(Z_1 Z_2) = E(Z_1)E(Z_2) = (0)(0) = 0$ eliminates the complicating middle term. Thus, *the variance of a sum of independent variables is the sum of the variances,* or the expression we have been using

$$\text{Var}(Y_n) = (\frac{1}{n})^2 \text{Var}(\sum X_j) = (\frac{1}{n})^2 \sum_j^n \text{Var}(X_j) = (\frac{1}{n})^2 n\sigma^2 = \frac{\sigma^2}{n}. \qquad (4.4)$$

Because $\sigma^2 = 0.96$, $\text{Var}(Y_n) = \frac{0.96}{n}$, which converts the example into

$$P\left( -\frac{0.10\sqrt{n}}{\sqrt{0.96}} \leq \frac{Y_n - E(Y_n)}{\sqrt{\text{Var}(Y_n)}} \leq \frac{0.10\sqrt{n}}{\sqrt{0.96}} \right). \qquad (4.5)$$

The values on the extreme sides determine the limits of integration for Equation 4.2. With $n = 100$, they are $a = -1.021$ and $b = 1.021$. By checking normal tables, the probability of the winnings being in this losing region is 0.69, which suggests that the likelihood of personal ruin is setting in.

Return to the "If I play just a bit more" mantra. More plays are represented by an increased value of $n$. Doing so reduces the variance, which means (Equation 4.5) that rather than a turn of luck, it becomes more likely that you will lose all money. Using $n$ values that are perfect squares, if $n = 20^2 = 400$, then $a = -2.041$ and $b = 2.041$ where the likelihood that the average loss is between $-.3$ and $-.1$ flirts with unity with the 0.958 value: Expect to go broke. The main point is that a decreasing variance makes it increasingly likely for the outcome to land closely around the mean.[1]

For our purposes, an important application of this observation is the following claim about $\Delta X \sim N(0, \Delta t)$.

**Proposition 1** *For* $\Delta X \sim N(0, \Delta t)$, *as the variance* $\Delta t \to 0$, *then with probability 1 the values of* $(\Delta X)^2$ *approach* $\Delta t$.

To indicate why this is so, if $Y = (\Delta X)^2$, then $E(Y) = E((\Delta X)^2)$. But $E(\Delta X) = 0$, so $E(Y) = E((\Delta X)^2) = \text{Var}(\Delta X) = \Delta t$. Coupled with the above intuition, it should be expected that, with increasing likelihood as $\Delta t \to 0$, the values allowed by $Y$ are very close to that of $\Delta t$.

This argument assumes that $\text{Var}(Y)$ becomes very small with $\Delta t$. While this true statement is, for now, taken on faith, the derivation of $Y$'s PDF starts in the next section.

### 4.1.1   Review of Chain Rule

The familiar chain rule, which simplifies computations such as where

$$\frac{d}{dx}(x^2 + 1)^{80} = 80(x^2 + 1)^{79}\frac{d}{dx}(x^2 + 1) = 160x(x^2 + 1)^{79},$$

is so commonly used that a full appreciation of its power can be lost. So, let's step back to review what the

$$\frac{d}{dt}f(g(t)) = f'(g(t))g'(t) \tag{4.6}$$

expression means. To visualize this equation, think of $z = f(x)$ as the altitude of a mountain in the East–West direction where $x$ denotes the distance. The $x = g(t)$ term determines a car's location at time t on this East–West scale. The composite equation $f(g(t))$ specifies the car's altitude at time t.

Clearly, the rate of altitude change is determined by the slope of the mountain *at the current location*, which is $f'(g(t))$, and how fast the car is moving, which is $g'(t)$. Thus Equation 4.6 captures the intuition that the rate of change of altitude is the product of two separate features: the mountain's slope and the car's speed.

This comment underscores a power of the chain rule; e.g., it can decompose economic factors. For instance, suppose your profit in manufacturing a particular product is based on the setting of a machine, $M$, where the speed of the raw product fed into the machine is determined by the setting of the assembly line, $A$. For added

---

[1] Once, after presenting this lesson, a student cried in anguish, "Why didn't you present this lesson *before* I went to Vegas last weekend!"

profit, should you deal with the machine or with the assembly line? To crudely examine this question, the profit equation is $P = M(A(t))$. According to the chain rule, the rate of change of profit is $\frac{dP}{dt} = M'(A(t))A'(t)$. The chain rule separates the efficiency of the machine, given by $M'$, from the effectiveness of the assembly line, given by $A'$, which provides tools for your analysis.

Returning to the chapter's theme, a second result from calculus is that an integral $\int_0^t f(x)\, dx$ can be computed with the antiderivative of $f(x)$. Namely,

$$F(t) - F(0) = \int_0^t f(x)\, dx, \quad \text{where } F'(t) = f(t). \tag{4.7}$$

Combining the fundamental theorem of calculus with the chain rule leads to a quick way to compute, say,

$$\frac{d}{dt}\left[ \int_{-t}^{t^2} e^{\sin(x)}\, dx \right]. \tag{4.8}$$

A horrendous approach would be to dutifully carry out all of the steps: First compute the integral (if you can!), substitute in the limits, and then differentiate. With the chain rule, everything becomes more civilized.

To do so with an abstract setting, suppose the goal is to compute

$$\frac{d}{dt}\left[ \int_{g(t)}^{h(t)} f(x)\, dx \right],$$

where $F(x)$ is the antiderivative of $f(x)$. This means that $\int_{g(t)}^{h(t)} f(x)\, dx = F(h(t)) - F(g(t))$. Because $F'(x) = f(x)$, it follows from the chain rule that

$$\frac{d}{dt}[F(h(t)) - F(g(t))] = F'(h(t))h'(t) - F'(g(t))g'(t) = f(h(t))h'(t) - f(g(t))g'(t).$$

Therefore,

$$\frac{d}{dt}\left[ \int_{g(t)}^{h(t)} f(x)\, dx \right] = f(h(t))h'(t) - f(g(t))g'(t). \tag{4.9}$$

What a powerful result! The problem can be resolved *without having to integrate anything!!* The barbarous chore of finding the antiderivative $F$ is completely avoided! The answer for Equation 4.8 is

$$\frac{d}{dt}\left[ \int_{-t}^{t^2} e^{\sin(x)}\, dx \right] = 2t e^{\sin(t^2)} + e^{\sin(-t)}.$$

### 4.1.2  Finding New PDFs

This change rule material introduces a way to derive PDFs that are needed in our finance discussions. To do so, recall that if $f(x)$ is the PDF for $X$, then the *cumulative distribution function*, or cdf, is given by

$$F(t) = P(X \le t) = \int_{-\infty}^{t} f(x)\, dx. \tag{4.10}$$

Using words, the cdf is the cumulative probability from $-\infty$ to t. Comparing Equation 4.10 with Equation 4.7 proves that the cdf is the antiderivative of the PDF.

This observation, combined with the chain rule, allows finding the PDF for different random variables. To illustrate with a special case, suppose $Y = X^2$ where the PDF for $X$ is $f(x)$. According to what was described (Equation 4.10), the PDF for $Y$ is $\frac{d}{dt} P(Y \le t)$. Therefore, to find $Y$'s PDF,

1. find an integral expression for $P(Y \le t)$,
2. then differentiate it.

For the first step, rewrite $Y$ in terms of what is known, which is $X$. So

$$P(Y \le t) = P(0 \le X^2 \le t) = P(-\sqrt{t} \le X \le \sqrt{t}) = \int_{-\sqrt{t}}^{\sqrt{t}} f(x)\, dx$$

is the sought after integral expression.

To find the PDF for $Y$, differentiate this integral by using Equation 4.9; the answer is

$$\frac{d}{dt} P(Y \le t) = \frac{d}{dt} \int_{-\sqrt{t}}^{\sqrt{t}} f(x)\, dx = \frac{1}{2} t^{-1/2} [f(\sqrt{t}) + f(-\sqrt{t})].$$

Using this approach, find the PDF for $Y = (\Delta X)^2$ and the Var$(Y)$.

For a simpler example, suppose you are in a contest where you must select a number at random from the interval $[0, 1]$. Your prize is determined by multiplying the selected number by 2 and then cubed. This means that the prizes range from 0 to 8. You want to compute the likelihood that you prize number will be between 1 and 4.

To handle this problem, let $X$ be the random variable of selecting a number at random from the interval $[0, 1]$, so its PDF $f(x) = 1$ for $0 \le x \le 1$. The prize value is given by $Y = (2X)^3 = 8X^3$, and the goal is to compute $Y's$ unknown PDF of $g(t)$. Whatever the form of $g$, it is clear that it equals zero for $t < 0$ and $t > 8$ (because the $Y = 8X^3$).

To find $g(t)$, compute

$$P(0 \leq Y < t) = P(0 \leq 8X^3 < t) = P(0 \leq X < \frac{1}{2}t^{\frac{1}{3}}) = \int_0^{\frac{1}{2}t^{\frac{1}{3}}} 1 \, dx.$$

Therefore the PDF for $Y$ is

$$g(t) = \frac{d}{dt} P(Y < t) = \frac{d}{dt} \int_0^{\frac{1}{2}t^{\frac{1}{3}}} dx = \frac{1}{6}t^{-\frac{2}{3}} \text{ for } 0 < t \leq 8.$$

The answer to the problem is

$$P(1 < Y < 4) = \frac{1}{6} \int_1^4 t^{-\frac{2}{3}} \, dt = \frac{1}{2}[4^{\frac{1}{3}} - 1].$$

## 4.2 Itô's Lemma

Beyond determining how prices will change, a more general goal is to understand how a financial function $f(S, t)$ changes. Following the lead of calculus, if $f(x) = x^2$, an approximation for the change (the derivative) is given by

$$\frac{\Delta f}{h} = \frac{f(x + h) - f(x)}{h} = \frac{(x + h)^2 - x^2}{h} = \frac{2xh + h^2}{h} = 2x + h,$$

which, when rewritten as a Taylor series approximation, becomes

$$\Delta f = 2xh + h^2.$$

The standard "take the limit as $h$ goes to zero" comment can be restated as "keep the slower, more dominant terms (multiples of $h$) and ignore terms that race off to zero much faster (the $h^2$ terms)." This leads to

$$\Delta f \approx 2xh.$$

This approach is used below by ignoring all terms smaller than $\Delta t$.

Itô's Lemma introduces a convenient method to approximate $\Delta f(S, t)$, which are changes in $f(S, t)$. Almost everything in finance revolves about some $f(S, t)$ choice. Already encountered are Puts (where $f(S, t) = P_E(S, t)$), Calls (where $f(S, t) = C_E(S, t)$), or even straddles (where $f(S, t) = P_E(S, t) + C_E(S, t)$). A key tool to understand the changing values of these financial instruments is Itô's Lemma.

**Theorem 2** *Assume that*

$$\Delta S = \sigma S \Delta X + \mu S \Delta t,$$

*where* $\Delta X \sim N(0, \Delta t)$. *If* $f(S, t)$ *is a smooth function, then for sufficiently small values of* $\Delta t$,

$$\Delta f(S, t) \approx \sigma S \frac{\partial f}{\partial S} \Delta X + [\mu S \frac{\partial f}{\partial S} + \frac{1}{2} \sigma^2 S^2 \frac{\partial^2 f}{\partial S^2} + \frac{\partial f}{\partial t}] \Delta t, \qquad (4.11)$$

*where the error tends to zero faster than* $\Delta t$.

Before indicating the proof, the critical thinking reader should be wondering why the mean for $\Delta X$ is zero and the variance equals $\Delta t$. The first is easy to answer: If the mean for $\Delta X$ is something else, move it to the drift term. As discussed in the previous chapter, the variance captures the random change over the $\Delta t$ time interval, so it is reasonable that the variance is some form of $\Delta t$. But why not $[\Delta t]^2$, so the standard deviation would be a multiple of $\Delta t$, or how about $\sqrt{\Delta t}$? Answers are important for the modeling, and so, keeping with this book's approach, the answer is embedded in Exercise 8.

Turning to the theorem, a way to appreciate Equation 4.11 is to compare it with the second order Taylor series expansion for a function $f(x, t)$ (e.g., see Equation 3.8) about the point $(x_0, t_0)$, which is

$$\Delta f = f(x, t) - f(x_0, t_0)) \approx \frac{\partial f}{\partial x}(x - x_0) + \frac{\partial f}{\partial t}(t - t_0) + \frac{1}{2!} \frac{\partial^2 f}{\partial x^2}(x - x_0)^2 + \\ + \frac{\partial^2 f}{\partial x \partial t}(x - x_0)(t - t_0) + \frac{1}{2!} \frac{\partial^2 f}{\partial t^2}(t - t_0)^2,$$

where all partial derivatives are evaluated at $(x_0, t_0)$. The error term tends to zero faster than the second order terms of $(t - t_0)^2$, $(x - x_0)(t - t_0)$, and $(x - x_0)^2$.

Replacing $x$ with $S$ yields the expression

$$\Delta f \approx \frac{\partial f}{\partial S} \Delta S + \frac{\partial f}{\partial t} \Delta t + \frac{1}{2} \frac{\partial^2 f}{\partial S^2}(\Delta S)^2 + \frac{\partial^2 f}{\partial S \partial t} \Delta S \Delta t + \frac{1}{2} \frac{\partial^2 f}{\partial t^2}(\Delta t)^2. \qquad (4.12)$$

Substituting $\Delta S$ into the first term on the right hand side of Equation 4.12 yields

$$\frac{\partial f}{\partial S} \Delta S = \sigma S \frac{\partial f}{\partial S} \Delta X + \mu S \frac{\partial f}{\partial S} \Delta t,$$

which accounts for all of the $\frac{\partial f}{\partial S}$ multiples in Equation 4.11. Similarly, the only Equation 4.11 term with a $\frac{\partial f}{\partial t}$ multiple is the second term on the right-hand side of Equation 4.12.

This comparison means that the remaining $\frac{1}{2} \sigma^2 S^2 \frac{\partial^2 f}{\partial S^2}$ term of Equation 4.11 represents *all* of Equation 4.12 second order terms. For this to happen, certain terms must drop out by being too small to be of any interest.

The elimination of minuscule terms is a standard mathematical approach, so it is worth developing intuition. Consider a driver's worries about a newly purchased Lamborghini when driving at unallowable speeds. If that well-polished surface is hit by a rock and a fly, the first is of concern, while that minute splat of the fly can be ignored. The same philosophy applies in mathematics: Terms of, say, $\Delta t \approx \frac{1}{100}$ can drive the modeling, but (at least for an approximation) those $(\Delta t)^2 \approx (\frac{1}{100})^2 = \frac{1}{10,000}$ terms are like that gutless fly; they are not worth considering.

The first casualty in dropping insignificant terms (those that go to zero faster than $\Delta t$) is the $(\Delta t)^2$ term of Equation 4.12 with $\frac{\partial^2 f}{\partial t^2}$. Similarly, as both $\Delta X$ and $\Delta t$ go to zero, $\Delta S \Delta t$ goes to zero faster than a speeding $\Delta t$, which allows ignoring the $\frac{\partial^2 f}{\partial S \partial t} \Delta S \Delta t$ term of Equation 4.12.

What remains is the $(\Delta S)^2 = \sigma^2 S^2 (\Delta X)^2 + 2\mu\sigma S^2 \Delta X \Delta t + \mu^2 S^2 (\Delta t)^2$ term. The expression's last two terms clearly go to zero faster than $\Delta t$, so it remains to examine $(\Delta X)^2$. This is where Proposition 1 plays a role; the proposition asserts that this random term is (with probability approaching one as $\Delta t$ approaches zero) approaching $\Delta t$. This term survives; replacing $(\Delta X)^2$ with $\Delta t$ completes the equation.

## 4.3 Application

Remember, a goal was to solve the expression

$$\Delta S = \sigma S \Delta X + \mu S \Delta t \qquad (4.13)$$

by finding the PDF for $S(t)$. To motivate what follows, suppose by magic it becomes possible to compute the precise future price of a commodity, perhaps an original edition album that a musical group called the "Rolling Stones" released with the song "(I Can't Get No) Satisfaction." Suppose the future value of the album's price, $S(t)$, (particularly at a desired time $t = T$) could be determined. Imagine the guaranteed profits that could result from such information!!

Unfortunately, this is not to be. But while finding the *exact* value of $S(t)$ is beyond current powers, with a couple of assumptions (that may or may not be valid), at least the PDF for $S(t)$ can be computed! Being armed with this PDF is powerful; it permits determining the likelihood that, for instance, on expiration date $t = T$ my Rolling Stones album is worth between \$110 and \$120. Finding the PDF for $S(t)$ provides a glimpse into the future. Not with a desired precision, but at least with a sense of likelihood.

This PDF is computed with Itô's expression Equation 4.11 where $f(S, t) = \ln(S)$. Why this choice? Well, express Equation 4.13 as

$$\frac{\Delta S}{S} = \mu \Delta t + \sigma \Delta X;$$

that term on the left-hand side appears to be $\Delta \ln(S)$.

As $f(S, t) = \ln(S)$ does not depend upon $t$, the $\frac{\partial f}{\partial t}$ term equals zero. Because $\frac{\partial f}{\partial S} = \frac{1}{S}$ and $\frac{\partial^2 f}{\partial S^2} = -\frac{1}{S^2}$, it follows, with a convenient cancelation, that

$$\Delta(\ln(S)) \approx \sigma \Delta X + (\mu - \frac{1}{2}\sigma^2)\Delta t. \tag{4.14}$$

An immediate application of Section 4.1.2 material (Exercise 3) proves that if $Y$ has a normal distribution, then so does $aY + b$. (More precisely, if $Y \sim N(\mu, \sigma^2)$, then $aY + b \sim N(a\mu + b, (a\sigma)^2)$. This makes sense; the $b$ merely translates the mean, while $a$ changes the scale.) Combining this fact with Equation 4.14 and the assumption that $\Delta X$ has a normal distribution leads to the important conclusion that $\Delta(\ln(S))$ *has a normal distribution*. It remains to find its mean and variance.

Computing the mean and variance is straightforward;

$$E(\Delta(\ln(S))) = \sigma E(\Delta X) + (\mu - \frac{1}{2}\sigma^2)\Delta t = (\mu - \frac{1}{2}\sigma^2)\Delta t.$$

Similarly, the variance is equal to the variance of $\sigma \Delta X$, so it is $\sigma^2(\Delta t)$.

Another straightforward application of Section 4.1.2 material shows that if $m$ and $s$ are, respectively, the mean and standard deviation of a normal distribution, then the PDF is

$$\frac{1}{\sqrt{2\pi}s}e^{-\frac{1}{2}(\frac{x-m}{s})^2}. \tag{4.15}$$

Equation 4.15 provides the PDF for $\Delta(\ln(S))$.

Fine, but the PDF for $\Delta(\ln(S))$ is *not* what is wanted: The goal is to find the PDF for $S(t)$. To compute this term, mimic the way in which integrals are defined: Divide the interval $[t_0, t]$ into $n$ equal subintervals with the notation $t_0 < t_1 < t_2 < \cdots < t_n = t$, $\Delta t = t_j - t_{j-1}$, and add Equation 4.14 increments. That is, compute both sides of

$$\sum_{j=1}^{n} \Delta(\ln(S)) \approx \sum_{j=1}^{n} \sigma \Delta X + \sum_{j=1}^{n} (\mu - \frac{1}{2}\sigma^2)\Delta t. \tag{4.16}$$

The sum on the left has the comfortable form

$$\ln(S(t)) - \ln(S(t_0)) = \sum_{j=1}^{n} \ln(S(t_j)) - \ln(S(t_{j-1})). \tag{4.17}$$

To explain, on the first $\Delta t = [t_0, t_1]$ interval, $\Delta \ln(S) = \ln S((t_1)) - \ln(S(t_0))$. Then on $\Delta t = [t_1, t_2]$, the increment is $\Delta \ln(S) = \ln(S(t_2)) - \ln(S(t_1))$. Adding these two terms together yields

$$[\ln(S(t_2)) - \ln(S(t_1))] + [\ln(S(t_1)) - \ln(S(t_0))] = \ln(S(t_2)) - \ln(S(t_0)),$$

because the two $\ln(S(t_1))$ values cancel. Adding the next increment $\ln(S(t_3)) - \ln(S(t_2))$ cancels the $\ln(S(t_2))$ terms leaving $\ln(S(t_3)) - \ln(S(t_0))$. In this manner, Equation 4.17 is obtained.

So far, the summation of the left-hand side of Equation 4.16 is determined. Summing the second component on the right-hand side of Equation 4.16 leads to $(\mu - \frac{1}{2}\sigma^2)(t - t_0)$.

What remains is the summation $\sum \sigma \Delta X$. Help comes from probability if we assume that what happens in each $\Delta t$ interval is independent of what happens in any other interval. (What an assumption! Is the reader willing to accept that if the price suddenly jumps in one time frame, it has no effect on what happens next?) With this strong condition, Equation 4.16 is the sum of independent random variables, each with a normal distribution, which allows invoking the following theorem.

**Theorem 3** *Let $X_1, \ldots, X_n$ be independent random variables where $X_j$ has a normal distribution with mean $m_j$ and variance $\sigma_j$, $j = 1, \ldots, n$. The random variable $Y = \sum_{j=1}^{n} X_j$ has a normal distribution with mean $m = \sum_{j=1}^{n} m_j$ and variance $\sigma^2 = \sum_{j=1}^{n} \sigma_j^2$.*

With this result, $\ln(S(t)) - \ln(S(t_0))$ has a normal distribution with mean

$$\sum (\mu - \frac{1}{2}\sigma^2)\Delta t = (\mu - \frac{1}{2}\sigma^2)(t - t_0).$$

By adding the constant term $\ln(S(t_0))$, the mean for $\ln(S(t))$ is

$$\ln(S(t_0)) + (\mu - \frac{1}{2}\sigma^2)(t - t_0).$$

Similarly, its variance is

$$\sum \sigma^2 \Delta t = \sigma^2(t - t_0).$$

Therefore,

**Theorem 4** *With all of the standard assumptions, the random variable $\ln(S(t))$ has a normal distribution with mean $\ln(S(t_0)) + (\mu - \frac{1}{2}\sigma^2)(t - t_0)$ and variance $\sigma^2(t - t_0)$.*

### 4.3.1  PDF for S(t)

Finally! Thanks to Theorem 4, the PDF for $S(t)$ can be computed. Finding the precise PDF is left as an exercise (Exercises 6aii, 14, and 15), which merely requires computing

$$\frac{d}{dx} P(S(T) \leq x).$$

Because $S(T)$ is the asset price at time $t = T$, $P(S(T) \leq x)$ is "the probability that the asset price at time $T$ is less than or equal to the value $x$."

Following the lead of Section 4.1.2, the first step is to find an integral expression for $P(S(T) \leq x)$. This is

$$P(S(T) \leq x) = P(\ln(S(T)) \leq \ln(x)) = \frac{1}{\sigma^* \sqrt{2\pi}} \int_{-\infty}^{\ln(x)} e^{\frac{(x-\mu^*)^2}{2(\sigma^*)^2}} \, dx, \qquad (4.18)$$

where $\sigma^*$ and $\mu^*$ have the values specified in Theorem 4. What remains is immediate: Differentiate! The resulting answer is called the *lognormal distribution*.

Once the PDF for $S$ is found, it becomes possible to compute the probability that, say, that Rolling Stones album will be worth between $100 and $120. Namely, if $g(x)$ is the PDF, then

$$P(100 < S < 120) = \int_{100}^{120} g(x) \, dx.$$

### 4.3.2  Lognormal Distribution

The central limit theorem captures a reason for the notable standing of the normal distribution. In this spirit, it is appropriate to add some words to underscore the special, widespread status of the lognormal distribution.

An explanation comes from the ubiquitous expression that is used to model change of a phenomenon $Y$ given by

$$\Delta Y = aY \Delta t + bY \Delta X, \quad \Delta X \sim N(0, \Delta t). \qquad (4.19)$$

Equation 3.20 is a special case for $\Delta S$. A reason why Equation 4.19 seems to be everywhere is that it simplifies a more accurate expression

$$\Delta Y = G(Y, \text{random effects}) \Delta t,$$

but where we may know absolutely nothing about $G$. This suggests adopting a Taylor series philosophy: Replace the mysterious $G$ with a linear approximation (i.e., Equation 4.19). For many settings, this first approximation of the unknown rate of growth, $\Delta Y$, depends on a multiple (the $a$ parameter, where $a = \mu$ in Equation 3.20) of the current size of $Y$. Similarly, the random effect, $\Delta X$, should be a multiple (the $b$ parameter, where $b = \sigma$ in Equation 3.20) of the $Y$. Whenever this is the case, and whenever Equation 4.19 occurs, expect the lognormal distribution to play a major role.

To appreciate where to anticipate such an expression, consider *almost anything* where amount of what is currently available can be expected to propagate the level of deterministic and random change. This comment immediately suggests ecology, with its issues about the growth and commonality spread among species. Preston [10] appears to be the first to discover that, indeed, the lognormal distribution is central for this analysis; a conclusion that has been strongly supported by subsequent ecological studies. Another example comes from communicable diseases; e.g., the more people who suffer the flu increases the risk that you will too. This is true whether the "communicable disease" is the infection time of online Twitter messages (e.g., [3, 8]), or even the dynamics of HIV [9].

Select just about any topic where the level of deterministic and random change depends on the size of the current status, and expect that the log-normal distribution plays a central role. This could be the growth of a fungus (Exercise 16), the acceptance of a particular technology, communication of information about the price of a commodity, or even the spread of a malicious but juicy rumor. Support for this assertion can be found with internet searches combining the object with lognormal, such as [rumors lognormal].

Why the lognormal? Several have wondered about the source of this ubiquitous distribution. Grönholm and Annila [5], for instance, articulated a commonly expressed concern: "Log-normal distributions describe data from diverse disciplines of science. However, the fundamental basis of log-normal distributions is unknown."

But it is known! An essential and somewhat satisfying answer is immediate: A first approximation for change of so many objects is modeled by Equation 4.19, which in turn ushers in the lognormal distribution.

This discussion also identifies when and why there are settings where the lognormal need not provide a good fit for data. After all, Equation 4.19 is a first approximation for change, which may not be adequate. A growing fungus, for example, may encounter a natural barrier, such as a pond or rocky region, which would require modifying at least the $bY\Delta Y$ term of Equation 4.19. A lognormal distribution is determined by its mean and variance (Exercise 15), but these values are determined by different features of the modeling (Theorem 4) that might change over time intervals; this could cause the graph of the lognormal PDF to vary. The modeling may require more than the first order Taylor series terms. But of value, we now know where and what to check.

Changing the modeling alters certain properties. As an illustration, Equation 4.19 is "scale-free." To suggest what this means, the change of variable $Y = 10U$ converts Equation 4.19 into

$$\Delta(10U) = a(10U)\Delta t + b(10U)\Delta X,$$

or the same expression

$$\Delta U = aU\Delta t + bU\Delta X.$$

But if Equation 4.19 must be modified to include a $Y^2$ term to have

$$\Delta Y = aY \Delta t + Y^2 \Delta t + bY \Delta X,$$

the change of scale expression differs; it now has the altered form

$$\Delta U = aU \Delta t + 10U^2 \Delta t + bU \Delta X.$$

The "scale-free" feature allows the expression to hold for behavior from the minuscule to the monumental. In particular, expect lognormal distributions to accompany such expressions. Conversely, if whatever is being modeled cannot enjoy such a scaling feature, then Equation 4.19 approximation needs to be re-evaluated.

## 4.4   Exercises

1. Find $\frac{d}{dt} \int_0^{e^t} e^{x^2} dx$.
2. The purpose of this exercise is to provide experience in finding PDFs.

   (a) Suppose $Y = 3X + 4$ where the PDF for $X$ equals 1 for $0 \le x \le 1$ and zero elsewhere. Find the PDF for $Y$.
   (b) Suppose $Y = X^2$ where the PDF for $X$ equals 1 for $0 \le x \le 1$ and zero elsewhere. Find the PDF for $Y$.
   (c) Suppose $f(x) = cx^2$ for $0 \le x \le 2$ and zero elsewhere is the PDF for $X$. Find the PDF for $Y = 3X + 4$.
   (d) Suppose $Y = X^4$ where $X \sim N(0, 4)$. Find the PDF for $Y$.

3. Suppose $X \sim N(0, 1)$ and $\sigma$ is a positive constant.

   (a) Find the PDF for $Y = \sigma X$.
   (b) Suppose $\mu$ is a constant; find the PDF for $Z = \sigma X + \mu$.

4. Let the PDF for $X$ be $f(x) = Cx$ for $0 \le x \le 1$ and zero elsewhere where $C$ is a constant you have to determine. Let $X(x) = x$.

   (a) Find the PDF for $Y = 2X + 4$.
   (b) Find the PDF for $Z = 4X^2$.
   (c) Let the PDF for $X(x) = x$ be given by $f(x) = e^{-x}$ for $x \ge 0$ and zero otherwise. Find the PDF for $Y = X^3$.

5. Suppose $X = \ln(Y)$ where $X \sim N(\mu, \sigma^2)$. Find the PDF for $Y$.
6. Assume that $X \sim N(\mu\sigma^2)$.

   (a) This exercise is to find the PDF for $S(T)$ as described in Theorem 4.

      i. Find the PDF for $Y = X^2$.

ii. Suppose $\ln(S(T)) \sim N(a, b^2)$. Find the PDF for $S(T)$. (To complete the above discussion concerning the distribution of $S$ with Theorem 4, $a = \ln(S(t_0)) + (\mu - \frac{1}{2}\sigma^2)(T - t_0)$ and $b^2 = \sigma^2(T - t_0)$.)

(b) Find the PDF for $Z = \exp(X)$.

(c) Find the PDF for $U = \ln(X)$.

(d) Find the PDF for $V = X^3$.

7. Suppose $\Delta X \sim N(0, \Delta t)$. Much of what was done uses the fact that $E((\Delta X)^2) = \Delta t$. By using the variance of $\Delta X$, show why this is so.

8. The next two problems explain why it is necessary to assume that the variance of $\Delta X$ is $\Delta t$. As it will become clear, should any other power be used, certain important terms would drop out of the equation.

(a) Suppose $\Delta X \sim N(0, (\Delta t)^2)$. Repeat the derivation of Itô's Lemma to find which terms disappear. That is, determine what terms are the largest when $\Delta t$ is very small, and keep only the terms of this magnitude. Explain why the answer is not satisfactory for our objectives.

(b) Redo part (a) but where the variance of $\Delta X$ is $\sqrt{\Delta t}$. Again, before canceling terms, check to see which terms have the largest value.

The next two problems indicate how to use Itô's Lemma for different choices of $\Delta S$.

9. If $\Delta S = .2S\Delta t + 2S\Delta X$, what is the form of Itô's Lemma?

10. Redo the above problem when $\Delta S = S^2\Delta t - S^3\Delta X$.

11. In the following, assume that $\Delta X \sim N(0, \Delta t)$ and then find the conclusion of Itô's Lemma.

(a) $\Delta S = 3S^2\Delta t + 6S^3\Delta X$.

(b) $\Delta S = \mu\Delta t + \sigma S\Delta X$.

(c) $\Delta S = 2S\Delta t + 3\sigma S^2\Delta X$.

(c) $\Delta S = \mu\frac{1}{S}\Delta t + \sigma S^2\Delta X$.

12. Let $X \sim N(0, 1)$.

(a) Let $Y = a + \sigma X$. Find the PDF for $Y$.

(b) Do the same for $Y = X^2$.

13. Suppose that $\Delta S = 3S\Delta t + 2S\Delta X$ and that today's price is $S(0) = 50$. *Find* (i.e., derive) the PDF for the price of $S(1)$, the price one year from now.

14. Complete the derivation of the PDF for $S(t)$; that is, find the derivative of Equation 4.18 and include the $\sigma^*$ and $\mu^*$ values.

15. These problems have to do with the lognormal distribution.

(a) The normal distribution satisfies

$$\frac{1}{\sigma\sqrt{2\pi}} \int_{-\infty}^{\infty} e^{-\frac{(x-\mu)^2}{2\sigma^2}} \, dx = 1.$$

Find the form of the integral after the substitution $x = \ln(s)$. This defines the PDF, $f(s)$, for the lognormal distribution where the limits of integration determine the allowed values of $s$.

(b) If $f(s)$ is the PDF found in the above problem, find where $f(s)$ has a maximum value. What happens to the location of this maximum value as $\sigma \to 0$? As $\sigma \to \infty$?

(c) From the information of the above two problems, and your knowledge of the shape of the normal distribution, give a rough sketch of the lognormal distribution for $\sigma = 0.1$ and $\mu = 2$. Do the same for $\sigma = 2$ and $\mu = 2$.

(d) It is clear from part a (and Exercise 14) that the PDF will have the form $e^{-\frac{(\ln(x)-\mu)^2}{2\sigma^2}}$. This suggests that a way to graph $e^{-\frac{(\ln(x)-\mu)^2}{2\sigma^2}}$ is to use $\ln(x)$ units on the horizontal axis. To indicate how to do so, represent $x$ as a power of a favored value $a > 1$, such as $x = 2^s$. With this choice, $\ln(x) = \ln(2^s) = s\ln(2)$. On the $s$-axis (the horizontal axis) write down the values with the normal, equally spaced coordinate ticks $\ldots, -2, -1, 0, 1, 2, 3, \ldots$; these will correspond to the $s$ or $\ln(x)$ values. Underneath each $s$ value, write down the corresponding $x = 2^s$ values of $\ldots, \frac{1}{4}, \frac{1}{2}, 1, 2, 4, 8, \ldots$. This means that one $s$ segment has all of the $x$ values between 1 and 2, the next equal sized $s$ segment has all $x$ values between 2 and 4, the following one has all values between 4 and 8, and so forth.

Draw the standard normal curve where $\mu = 2$ is in terms of $x$. This provides an expanded display of the lognormal distribution.

Next, to obtain a usual depiction, draw the graph obtained by rescaling the $x$ values so they are equally spaced. This leads to a serious contraction from 0 to 1 (everything from $-\infty$ to 0 on the $s$ scale must be squeezed into a unit interval of the $x$ scale) accompanied by a massive expansion beyond 1 (e.g., everything from $0 < s < 3$ on the $s$-scale must be expanded to the 1 to 8 units on the $x$-scale).

16. In Crystal Falls, Michigan (located in the Upper Peninsula of Michigan— Michigan consists of two disconnected peninsulas; the Upper Peninsula is the northern one with shores on Lake Superior), an enormous fungus, with radius of 600 feet, is growing. Suppose the growth of the circumference, $C(t)$, is such that

$$\Delta C(t) = 0.1 C(t)\Delta t + 0.2 C(t)\Delta X, \quad \Delta X \sim N(0, \Delta t). \qquad (4.20)$$

Find the probably distribution for $C(100)$ (the circumference a century from now) and state the assumptions needed to derive this statement. Are the assumptions reasonable?

17. This exercise provides an opportunity to use Itô's Lemma. Assume that $\Delta S = \mu S \Delta t + \sigma S \Delta X$ with $\Delta X \sim N(0, \Delta t)$.

 (a)  Find $\Delta \ln(S)$.

(b)  Find $\Delta[[C_E(S, t)]^2]$.

(c)  Find $\Delta[C_E(S, t) - 4S]$.

(d)  Find $\Delta[P_E(S, t) + 2S]$.

(e)  Find $\Delta[C_E(S, t) - P_E(S, t)]$.

# Chapter 5
# The Black–Scholes Equation

We now are ready to derive the important Black–Scholes Equation [1], which is widely used to determine pricing of Calls and Puts! An outline is given next; details are developed in the next chapter.

1. *Hedging!* The Black–Scholes Equation describes an interesting double hedge: The first hedge involves a portfolio's market structure; the second is a hedge between markets.

   More precisely, suppose a portfolio $\Pi$ is carefully hedged with Puts, Calls, going long and short, to ensure against disaster. But there is nothing in the portfolio's design that requires an investor to stay in one market. Should opportunities arise elsewhere, it is reasonable to transfer money to a competing financial opportunity. One of the several possibilities, a neutral one, is to sell $\Pi$ and invest the money in a bank.

2. *Itô's Lemma.* Investors follow the change in investments. So, with the described options, compute $\Delta_{market}\Pi$ and $\Delta_{bank}\Pi$. The second term (as developed in Section 2.4.1) with interest rate $r$ is

$$\Delta_{bank}\Pi = r\,\Pi\,\Delta t. \tag{5.1}$$

   To compute $\Delta_{market}\Pi$, Itô's Lemma assumes a central role.

3. *Arbitrage.* Differences between $\Delta_{market}\Pi$ and $\Delta_{bank}\Pi$ can create arbitrage opportunities. As investors can be expected to move in the direction of advantage, the accompanying changes in prices, interest rates, and other market features lead to $\Delta_{market}\Pi = \Delta_{bank}\Pi$.

4. *Football gambles, eliminating risk.* Explicit risk is introduced by the random term $\Delta X$. Similar to how risk in the football examples is reduced by adjusting bets, this random $\Delta X$ risk is eliminated by adjusting the portfolio.

© Springer Nature Switzerland AG 2019

D. G. Saari, *Mathematics of Finance*, Undergraduate Texts in Mathematics,

https://doi.org/10.1007/978-3-030-25443-8_5

5. *Put–Call Parity.* Several technical details are resolved by appealing to the Put–Call Parity Equation.

With this guideline, let us move to the derivation.

## 5.1  Black–Scholes Equation

To appreciate the implications and power of Itô's Lemma,

$$\Delta f(S, t) = \sigma S \frac{\partial f}{\partial S} \Delta X + \left[ \mu S \frac{\partial f}{\partial S} + \frac{1}{2} \sigma^2 S^2 \frac{\partial^2 f}{\partial S^2} + \frac{\partial f}{\partial t} \right] \Delta t, \qquad (5.2)$$

let $V(S, t)$ be the value of an option, perhaps Puts or Calls with expiration price $E$ at expiration date $T$. The goal is to find the current value of $V(S, t)$ for any asset price $S$ value at any time $t$ prior to the expiration date.

To recognize why we should do so, suppose Torik owns a Call for E = \$100 where T is March 1 and today's price is S = \$102. Tatjana can buy this Call for \$3.70; is this a good or bad deal? An answer would follow immediately if the market value of $C_{100}(102, t)$ could be computed.

Remember, options are not bought for amusement; they are parts of investments to make money. Without allegiance to any particular stock, the value of an option reflects a balance between owning the asset, hedging properties, and alternative ways to earn money through other investments such as placing the money in the bank.

To start, assume that a portfolio consists of a combination of the option and $\delta$ units of the asset. The purpose of $\delta$ is to find an appropriate hedge mixture between the amounts of options and an asset. With football bets, this is akin to betting \$100 with Bob and then determining how much to bet with Sue to eliminate risk and ensure a fixed return. The $\delta$ term, called the "hedge ratio," plays a similar role; its value during any $\Delta t$ period of time *must* be determined because it identifies how to adjust a portfolio to reduce risk. The next $\Delta t$ time interval can be compared with a new football game, so expect the $\delta$ value to change.[1]

The portfolio is modeled as

$$\Pi(S, t) = V(S, t) - \delta S. \qquad (5.3)$$

The fact $\delta$ is a constant leads to the relationships

$$\frac{\partial \Pi}{\partial S} = \frac{\partial V}{\partial S} - \delta, \quad \frac{\partial^2 \Pi}{\partial S^2} = \frac{\partial^2 V}{\partial S^2}, \quad \frac{\partial \Pi}{\partial t} = \frac{\partial V}{\partial t}.$$

---

[1]Notation change! Although the literature usually uses *Delta*, in this book $\Delta$ represents mathematical change. So, for this hedging purpose, use the lower case $\delta$.

Substituting these values into Itô's Lemma defines

$$\Delta_{market} \Pi(S, t) = \left(\frac{\partial V}{\partial S} - \delta\right)\sigma S \Delta X + \left[\left(\frac{\partial V}{\partial S} - \delta\right)\mu S + \frac{1}{2}\sigma^2 S^2 \frac{\partial^2 V}{\partial S^2} + \frac{\partial V}{\partial t}\right]\Delta t.$$
(5.4)

Remember, the purpose of the hedge ratio $\delta$ is to remove risky consequences caused by random $\Delta X$ changes. A natural way to eliminate these random terms is to set the coefficient of the $\Delta X$ term equal to zero. Doing so defines the value

$$\delta = \frac{\partial V}{\partial S}.$$
(5.5)

This is done at the beginning of each $\Delta t$; hold this value constant throughout the time interval. In terms of the football wager, should Bob change his odds, the amount wagered with Sue should also change. Similarly, expect the $\delta$ value, which determines the composition of the portfolio, to change with differing events, such as the values of $V$ and $S$, in the next time interval.

To appreciate this $\delta$ value, the *hedge ratio*, recall that (by definition) $\frac{\partial V}{\partial S}$ measures how $V$'s value changes with respect to the price $S$. For this reason, $\delta$ identifies how to readjust a portfolio to achieve the desired level of hedging. Indeed, the sign of $\delta = \frac{\partial V}{\partial S}$ reflects the different ways to manage a portfolio: When $\delta$ is positive, the $-\delta S$ term requires going short; when $\delta$ is negative, the $-\delta > 0$ value requires going long.

As an intuition pause, what should be the sign of $\delta = \frac{\partial C_E(S,t)}{\partial S}$? For $\delta = \frac{\partial P_E(S,t)}{\partial S}$? To answer this question, recall what causes $C_E(S, t)$, or $P_E(S, t)$, to increase (or decrease) in value.

A consequence of Equation 5.5 risk management choice of the hedge ratio $\delta$ is that *two terms*—not just the expression with $\Delta X$—drop out of Equation 5.4. This reduces the expression to the much simpler

$$\Delta_{market} \Pi(S, t) = \left[\frac{1}{2}\sigma^2 S^2 \frac{\partial^2 V}{\partial S^2} + \frac{\partial V}{\partial t}\right]\Delta t.$$
(5.6)

An interesting surprise is that Equation 5.6 has no $\mu$ terms. That is, a consequence of removing risky random effects is to also eliminate drift behavior reflected by the parameter $\mu$. This means that, presumably, two traders with different estimates of $\mu$ could reach the same conclusion about the value of $V$. What remains are volatility effects captured by $\sigma$.

So far, Equation 5.6, along with the hedge ratio, represents the value of option $V$ as affected by the asset's current price. But financial openings are not restricted to a single asset; be honest, in your own finances, say in purchasing a new car, if you see a better offer elsewhere, you probably would go for it. This requires comparing the value of option $V$ with other market opportunities. General market effects reflect changes in the value of money—the interest rate. So the next step is to compare changes in the value of an option with that of the value of money.

The notion to be used is that if the current rate of interest in a risk free loan, or a deposit in the bank, is $r$, then the amount of money made over interval $\Delta t$ on an investment of $\$\Pi$ is

$$\Delta_{bank}\Pi = r\Pi\Delta t. \tag{5.7}$$

But how is $\Delta_{bank}\Pi$ from Equation 5.7 related to $\Delta_{market}\Pi$ in Equation 5.6?

Suppose in a cost free manner, a portfolio could be converted into money at any desired time, or, in the opposite direction, money could be borrowed at the interest rate $r$ to buy more options and/or assets. If changes in the portfolio are more favorable than what money costs to borrow, then borrow money from your friendly banker to increase the portfolio holdings. This arbitrage opportunity makes money more valuable, so, presumably, the interest rate $r$ should go up (yes, another assumption), which eventually squeezes out advantages of borrowing money.

Conversely, if more money can be made with a bank deposit at the $r$ rate of interest, sell off a portion of a portfolio and invest the money in a bank. Should many people do so, expect the interest rate, $r$, to decrease.

This reasonable argument has a seeming flaw. Because $S$ can be any asset, including Rolling Stones albums, it requires a huge stretch of imagination to accept that the interest rate, $r$, can be influenced by sales of Rolling Stones albums! Of course not! But the market is interconnected, so the value of the albums are related to "whatever," which are connected to "something else," and "even other stuff." So, rather than the neutral choice of a bank, expect the value of the portfolio $\Pi(S, t) = V(S, t) - S\frac{\partial V}{\partial S}$ to react to the general market. The appropriate arbitrage argument is left for the reader to carry out.

Combining the *arbitrage* arguments, in a highly idealized setting where there is an instantaneous change in interest rates and total understanding of what course of action is most profitable, it is reasonable to expect that $\Delta_{market}\Pi = \Delta_{bank}\Pi$. Setting them equal and using Equations 5.3, 5.5 leads to

$$r(V - S\frac{\partial V}{\partial S})\Delta t = [\frac{1}{2}\sigma^2 S^2 \frac{\partial^2 V}{\partial S^2} + \frac{\partial V}{\partial t}]\Delta t.$$

Dividing both sides by $\Delta t$ leads to a simple version of the *Black–Scholes Equation*

$$\frac{\partial V}{\partial t} + \frac{1}{2}\sigma^2 S^2 \frac{\partial^2 V}{\partial S^2} + rS\frac{\partial V}{\partial S} - rV = 0. \tag{5.8}$$

A solution of this equation determines the value of a Put, Call, or any combination, for any specified $S$ and $t < T$. Wait! How can this assertion be true? After all, a Call's value is based on increasing prices; a Put is based on an expectation of decreasing prices. How can the same equation bundle opposing directions? Something is missing. There is, and it is determined next.

## 5.2 Boundary Conditions

The Black–Scholes Equation is a *backwards parabolic equation*. The term parabolic comes from comparing the highest order partial derivatives of both variables with quadratic equations. Recall, a *parabolic equation* has the

$$y = ax^2 + bx + c \tag{5.9}$$

structure, where $y$ is raised to the first power and the highest $x$ power is squared. Because Equation 5.8 has the same form, where derivatives replace the powers, it is called a *parabolic partial differential equation.*

To describe the "backwards" term, retaining only the highest order derivatives terms leads to

$$\frac{\partial V}{\partial t} = -\frac{1}{2}\sigma^2 S^2 \frac{\partial^2 V}{\partial S^2}. \tag{5.10}$$

For the moment, concentrate on the negative sign. To appreciate the impact of this sign, consider a simple differential equation

$$\frac{dy}{dt} = -5, \quad y(t_0) = 1$$

with the unique solution

$$y(t) - 1 = -5(t - t_0),$$

where the *negative* multiple of the time term indicates a "backward" moving solution. Similarly, solutions for the Black–Scholes Equation move backwards from the expiration time of $T$. Actually, this motion is precisely what is desired; the goal is to unravel what happens *before* the expiration date!

### 5.2.1  Heat Equation

To identify what information is missing from the Black–Scholes Equation, turn to the *forward* parabolic partial differential equation given by

$$\frac{\partial u(x, t)}{\partial t} = \frac{\partial^2 u(x, t)}{\partial x^2}. \tag{5.11}$$

Equation 5.11 is the *heat equation*. As an example, consider a thin bar that is 20 units long. The solution for the heat equation, $u(x, t)$, identifies the temperature of the bar at each $x$ location, $0 \leq x \leq 20$, at any instant of time $t \geq 0$. But finding the answer requires supplementary information.

For instance, what was the initial status of the bar? Was it taken out of a freezer, a furnace? Different choices must lead to contrasting $u(x, t)$ solutions. Thus, one type of information is the bar's initial status given by

$$u(x, 0) = u_0(x), \tag{5.12}$$

where function $u_0(x)$ specifies the bar's initial temperature at each position. As an example, if the temperature is uniformly distributed ranging from $0^o$ at $x = 0$ to $100^o$ at $x = 20$, then

$$u(x, 0) = \frac{100x}{20}. \tag{5.13}$$

Even should two bars have identical initial temperature distributions, expect mismatched outcomes should the ends of the bars be treated differently for $t > 0$. To illustrate this comment, consider dissimilar heat distributions for three bars, where applying a blow-torch to each end of the first bar, a blow-torch is heating one end but ice tied to the other end of a second bar, and ice is applied to both ends of the third bar will cause differing answers. This means that information about $u(0, t)$ and $u(20, t)$ also is needed. The choice of keeping both ends at zero temperature forever after the start defines the conditions

$$u(0, t) = 0, \quad u(20, t) = 0. \tag{5.14}$$

All of this can be depicted in an infinite rectangular region

$$\{(x, t) \mid 0 \leq x \leq 20, \quad t \geq 0\}$$

with its three boundaries: The horizontal line segment $(x, 0)$, $0 \leq x \leq 20$ on the $x$-axis, and two boundaries that are infinitely long corresponding to $(x = 0, t \geq 0)$ and $(x = 20, t \geq 0)$. A solution requires specifying the desired behavior of $u(x, t)$ along each boundary segment.

### 5.2.2    Black–Scholes Boundary Conditions

Guided by the heat equation, it becomes clear that the values of option $V$ along the boundaries must be specified. But, what boundaries?

Rather than describing $u(x, t)$ at starting time $t = 0$, with a backward equation, it is the value of $V$ at expiration date $t = T$ that must be specified. Then, mimicking expressions for $u(x, t)$ at endpoints of the bar, what is needed are the $V$ values at the price extremes of $S = 0$ and where $S \to \infty$. This means that the values of an option $V$ occur in the region

$$\{ (S, t) \,|\, 0 \leq S < \infty, \quad -\infty < t \leq T \}. \tag{5.15}$$

## Call Conditions

As true with the heat equation, the values of option $V$ along the boundaries of Equation 5.15 region must be specified. To be specific, let $V(S, t) = C_E(S, t)$; this is a Call with strike price $E$ at expiration date $T$.

The value of $C_E(S, t)$ along the $t = T$ boundary was developed in Chapter 2; according to Equation 2.1, it is

$$C_E(S, T) = \max(S - E, 0). \tag{5.16}$$

Along the zero price boundary $S = 0$, it is reasonable to accept that $C_E(0, t) = 0$. After all, according to the price change equation,

$$\Delta S = \sigma S \Delta X + \mu S \Delta t,$$

once $S = 0$, it remains zero for all time. (Substituting $S = 0$ into the right-hand side of this expression shows that a subsequent change in value, determined by $\Delta S = 0$, is zero.) This defines the second boundary condition

$$C_E(0, t) = 0. \tag{5.17}$$

It remains to find the boundary conditions for $S \to \infty$. It is reasonable to expect that $C_E(S, t) = S - E$, where, with an infinite value for the asset price, the $E$ value becomes insignificant. Consequently, the value of $C_E(S, t)$ is, essentially, $S$. A worry is whether an infinitely large price might suddenly collapse. The essence of the argument showing this is not a problem is outlined in a footnote.[2] The final condition is

$$C_E(S, t) \to S \text{ as } S \to \infty. \tag{5.18}$$

---

[2] Start with $Y = 1/S$, and use Itô's Lemma to derive $\Delta Y = -\sigma Y \Delta X + [-\mu Y + \sigma^2 Y]\Delta t$. Assuming that $S$ has a very large value is essentially the same as assuming that $Y = 0$. However, the equation for $\Delta Y$ is such that $Y = 0$ requires $Y$ to remain zero for all time. In turn, $S$ remains infinitely large for all time.

## Put Conditions

For a Put, money is made only when the asset price decreases. And so, as developed
for Equation 2.3, the boundary condition for $t = T$ is

$$P_E(S, T) = \max(E - S, 0). \tag{5.19}$$

Similarly, should the asset price grow without bound, the Put becomes useless. Thus,

$$P_E(S, t) = 0 \text{ as } S \to \infty. \tag{5.20}$$

The final boundary condition is where the asset price drops to zero. The earlier
argument using the $\Delta S$ structure ensures that the asset price will not change, which
means at time $t = T$ that $P_E(0, T) = E$. At an earlier time, the value of the option
is the present value of $E$.

To compute this present value of $E$, use the differential equation

$$\frac{dm}{dt} = rm, \quad m(T) = E,$$

where $m(t)$ represents the money. The solution is

$$\int_{m(t)}^{E} \frac{1}{s} ds = \int_{t}^{T} dt, \quad m(t) = Ee^{-r(T-t)}.$$

Therefore, the last boundary condition for a Put is

$$P_E(0, t) = Ee^{-r(T-t)}. \tag{5.21}$$

## Help from an Old Friend

The inclusion of the present value of $E$ for $P_E(0, t)$ (Equation 5.21) may seem
mysterious. An explanation comes from our old friend the Put–Call Parity Equation

$$P_E(S, t) + S = C_E(S, t) + Ee^{-r(T-t)}.$$

By substituting $S = 0$, $C_E(0, t) = 0$ into this expression, the desired $P_E(0, t) =$
$Ee^{-r(T-t)}$ boundary value emerges. Later, after finding the $C_E(S, t)$ solution, the
Put–Call Parity Equation will be used to obtain the $P_E(S, t)$ solution.

**Comparisons**

The *same* Black–Scholes Equation governs the values of a Put *and* a Call. What differs are the boundary conditions. The same reason why different boundary choices for the heat equation can lead to radically different solutions explains why the different boundary conditions for a Put, or a Call, or some combination (which would require different boundary conditions) create different behaviors.

The solutions of these equations provide valued information: They specify the value of an option at any time $t$ for any asset price $S$. This means that when entering the market, you know what to expect. What remains is to indicate how to solve the Black–Scholes Equation.

## 5.3 Conversion to the Heat Equation

In its current form, the Black–Scholes Equation is difficult to solve. But by using the time honored mathematical approach of changing variables to create a simpler problem, the equation is converted into the heat equation.

To indicate why it is of interest to describe the change of variables in a discussion of the *mathematics* of finance, compare it with solving $\int \cot(x)\, dx$. Rewriting this expression as $\int \frac{\cos(x)}{\sin(x)}\, dx$ identifies the $\sin(x)$ denominator as the ugly (or $u$) part in a change of variables. Setting $u = \sin(x)$ leads to the simpler $\int \frac{1}{u}\, du = \ln(|u|) + C$ expression with the solution $\ln(|\sin(x)|) + C$. Now, with integral tables readily available on the internet, the solution can be easily found. But to appreciate why the $\sin(x)$ term is in the answer, the change of variable must be understood. Similarly, to recognize the source of terms in solutions of the Black–Scholes Equation, the change of variables used to convert the equation to the heat equation must be tracked.

A clue how to solve the Black–Scholes Equation comes from the similarity of Equation 5.10 with the heat equation, Equation 5.11. OK, the first is a backward parabolic equation, while the second is forward, but a change of the independent variable $t = -\tau$ handles that difference. A more serious concern is the variable coefficient of Equation 5.10, while the heat equation has constant coefficients.

## *5.3.1 A Quick Tutorial in Differential Equations*

An important mathematical tool, which is evoked to justify replacing variable coefficients with constant ones, is the existence and uniqueness theorems. The value of these results is that they identify which differential equations can be solved and have unique solutions. Stated differently, this powerful assertion provides a license

to find a solution in any desired way; if a wild guess turns out to be a solution, that
is fine.

Start with

$$\frac{dy}{dx} = 3y. \tag{5.22}$$

As described earlier (e.g., when computing the present value of money), an answer
is given by

$$\frac{dy}{y} = 3dx, \quad \ln(y) = 3x + c_1$$

or $e^{\ln(y)} = e^{3x+c_1} = e^{3x}e^{c_1} = Ce^{3x}$. As $e^{\ln(y)} = y$, the solution is $y = Ce^{3x}$.
The value of $C$ is determined by initial conditions. If $y(0) = 5$, then $5 = y(0) = Ce^{3\times0} = C$, so the solution is $y(x) = 5e^{3x}$.

Now consider a more difficult problem

$$y'' - 5y' + 6y = 0 \tag{5.23}$$

with initial conditions $y(0) = 1$, $y'(0) = 4$. Knowing that a unique solution exists
provides a license to find an answer by guess or trial and error.

A reasonable first choice is to mimic Equation 5.22 solution of $y(x) = Ce^{mx}$ for
some value of $m$. Ignore the $C$ for now (it is a constant of integration), and seek the
value of $m$. That is, if $y = e^{mx}$ is a solution, then

$$y = e^{mx}, \quad y' = me^{mx}, \quad y'' = m^2 e^{mx}.$$

Substituting this hunch into Equation 5.23 leads to

$$m^2 e^{mx} - 5me^{mx} + 6e^{mx} = e^{mx}[m^2 - 5m + 6] = 0.$$

Consequently, the task of solving the differential equation reduces to the high school
problem of finding the roots of the equation

$$m^2 - 5m + 6 = 0, \quad (m - 3)(m - 2) = 0, \quad m = 2, 3.$$

It now follows that the solution is

$$y(x) = C_1 e^{3x} + C_2 e^{2x}.$$

It remains to find the values of $C_1$ and $C_2$ from the given information that

$$1 = y(0) = C_1 e^0 + C_2 e^0 = C_1 + C_2,$$

and

$$4 = y'(0) = 3C_1e^0 + 2C_2e^0 = 3C_1 + 2C_2.$$

The solution of this "two equations, two unknowns" system is $C_1 = 2, C_2 = -1$, so the answer for Equation 5.23 is $y(x) = 2e^{3x} - e^{2x}$.

Now a twist. To indicate how to remove the $S^2$ and $S$ variable coefficients of the Black–Scholes Equation, consider the problem of solving

$$x\frac{dy}{dx} = 3y. \tag{5.24}$$

The change-of-variable plan is to replace the variable $x$ with some $u(x)$.

Should the mathematical universe be accommodating, Equation 5.24 could be converted into an Equation 5.22 form. According to the chain rule, any choice of $u$ leads to

$$\frac{dy}{dx} = \frac{dy}{du}\frac{du}{dx}.$$

While it is unclear how to select $u(x)$, a desired choice would lead to

$$x\frac{dy}{dx} = x\left[\frac{dy}{du}\right]\left[\frac{du}{dx}\right] = 3y.$$

Aha! Selecting $u(x)$ so that $x\left[\frac{du}{dx}\right] = 1$ defines the equation

$$x\left[\frac{du}{dx}\right]\frac{dy}{du} = \frac{dy}{du} = 3y, \tag{5.25}$$

which is Equation 5.22 with a different independent variable. Therefore the general solution is

$$y = Ce^{3u}. \tag{5.26}$$

The choice of a $u(x)$ that generates the magic whereby $x\left[\frac{du}{dx}\right] = 1$ requires $\frac{du}{dx} = \frac{1}{x}$. This means that $u(x) = \ln(x)$, so $u(x) = \ln(x)$ *is a desired change of variables!* With this choice, Equation 5.26 becomes

$$y = Ce^{3u} = Ce^{3\ln(x)} = Ce^{\ln(x^3)} = Cx^3.$$

The Black–Scholes Equation has $S^2$ and $S$ coefficients, which suggests understanding how to handle equations of the form

$$x^2y'' - 4xy' + 6y = 0. \tag{5.27}$$

The goal is to convert Equation 5.27 type of equations into Equation 5.23 format of constant coefficient equations that can be solved.

Success was achieved with $u(x) = \ln(x)$ for first order equations, where, so far, the change of variables defines

$$\frac{dy}{dx} = \frac{dy}{du}\frac{du}{dx} = \frac{1}{x}\frac{dy}{du}. \tag{5.28}$$

Checking what happens with the second derivative leads to

$$y'' = \frac{d}{dx}\left[\frac{dy}{dx}\right] = \frac{d}{dx}\left[\frac{1}{x}\frac{dy}{du}\right] = -\frac{1}{x^2}\frac{dy}{du} + \frac{1}{x}\frac{d}{dx}\left[\frac{dy}{du}\right] =$$

$$-\frac{1}{x^2}\frac{dy}{du} + \frac{1}{x}\frac{d}{du}\left[\frac{dy}{du}\right]\frac{du}{dx}, \tag{5.29}$$

or, because $\frac{du}{dx} = \frac{1}{x}$,

$$y'' = -\frac{1}{x^2}\frac{dy}{du} + \frac{1}{x^2}\frac{d^2y}{du^2}. \tag{5.30}$$

The first part of Equation 5.29 is the product rule, and the last part is the chain rule. (Remember, the chain rule has

$$\frac{d(\ )}{dx} = \frac{d(\ )}{du}\left[\frac{du}{dx}\right]$$

for whatever is in the ( ) brackets. Here, the bracket term is $\frac{dy}{du}$.)

Replacing $y'$ with Equation 5.28 and $y''$ with Equation 5.30 converts Equation 5.27 into

$$x^2\left[-\frac{1}{x^2}\frac{dy}{du} + \frac{1}{x^2}\frac{d^2y}{du^2}\right] - 4x\left[\frac{1}{x}\frac{dy}{du}\right] + 6y = \frac{d^2y}{du^2} - 5\frac{dy}{du} + 6y = 0. \tag{5.31}$$

There is magic!! Equation 5.31 agrees with Equation 5.27 with the independent variable $u$ rather than $x$. Equation 5.27 has been solved, so Equation 5.31 solution is

$$y(u) = C_1 e^{3u} + C_2 e^{2u}.$$

Changing $u$'s name to $x$ leads to

$$y(x) = C_1 e^{3\ln(x)} + C_2 e^{2\ln(x)} = C_1 e^{\ln(x^3)} + C_2 e^{\ln(x^2)} = C_1 x^3 + C_2 x^2.$$

## 5.3.2 Eliminating the Variable Coefficients

The first step in converting the Black–Scholes Equation into an easier expression is to eliminate the $S$ coefficients. This is done the same manner as with differential equations.

To find an appropriate change of variables $x = x(S)$, recall from the chain rule that

$$\frac{\partial V}{\partial S} = \frac{\partial V}{\partial x} \frac{dx(S)}{dS}.$$

To meet the goal of eliminating the $S$ multiple of $rS\frac{\partial V}{\partial S}$, define

$$\frac{dx(S)}{dS} = \frac{1}{S},$$

or

$$x(S) = \ln(S) + \ln(c) = \ln(cS), \tag{5.32}$$

where $c$ is a constant that is selected to simplify the boundary conditions.

This change of variable leads to

$$\frac{\partial V}{\partial S} = \frac{1}{S}\frac{\partial V}{\partial x}, \quad \frac{\partial^2 V}{\partial S^2} = \frac{\partial}{\partial S}\frac{1}{S}\frac{\partial V}{\partial x} = \frac{1}{S^2}\left[\frac{\partial^2 V}{\partial x^2} - \frac{\partial V}{\partial x}\right].$$

Substituting these values into the Black–Scholes Equation 5.8 leads to

$$\frac{\partial V}{\partial t} + \frac{\sigma^2}{2}\frac{\partial^2 V}{\partial x^2} + (r - \frac{\sigma^2}{2})\frac{\partial V}{\partial x} - rV = 0. \tag{5.33}$$

Notice the coefficients: It is reasonable to anticipate that these terms will appear as parts of the Black–Scholes solution. This is the case.

This change also affects the boundary conditions. Express this change as $cS = e^x$ where $c = \frac{1}{E}$. This leads to

$$e^x = \frac{S}{E}, \text{ or } x(S) = \ln\left(\frac{S}{E}\right), \tag{5.34}$$

so the boundary conditions at time $t = T$

$$C_E(S, T) = \max(S - E, 0) = E\max(S/E - 1, 0)$$

become

$$E \max(e^x - 1, 0).$$

To eliminate the $E$ multiple of the max term, let $V = Ev$, so that

$$\frac{\partial v}{\partial t} + \frac{\sigma^2}{2} \frac{\partial^2 v}{\partial x^2} + (r - \frac{\sigma^2}{2}) \frac{\partial v}{\partial x} - rv = 0 \qquad (5.35)$$

with initial conditions

$$v(x, T) = \max(e^x - 1, 0). \qquad (5.36)$$

As for the other boundary conditions, as $S \to 0$, $x \to -\infty$, so the $C_E(0, t) = 0$ condition translates into $v(x, t) \to 0$ as $x \to -\infty$. Similarly, the condition that $C_E(S, t) \to S$ as $S \to \infty$ becomes $v(x, t) \to S/E = e^x$ as $x \to \infty$.

A next step is to eliminate the $\frac{\partial v}{\partial x}$ and $v$ terms.[3] But this change does not introduce terms that are needed to appreciate Black–Scholes solutions, so it is not carried out.

## 5.4   Intuition

At this stage, the reader is encouraged to speculate about what terms to anticipate in solutions for the Black–Scholes Equation. It is reasonable to expect $r - \frac{\sigma^2}{2}$ expressions, because they cropped up with the change of variables. Anything else? An important change of variables involved $\ln(S/E)$, so we should expect that this natural log expression will assume a dominant role; it does. Investing even a small amount of time to develop these conjectures is a strong way to better understand an area.

It also is worth exploring what might happen with changes in the market. Suppose the market is very volatile as captured by a large $\sigma$ value. How should this affect the value of $C_E(S, t)$? Of $P_E(S, t)$? What about a stagnant market where $\sigma$ has a small value? An important market variable is the interest rate $r$; how would increasing and decreasing $r$ values affect the options? What about if $\sigma$ decreases but $r$ increases?

---

[3] The standard approach, which mimics completing the square, is to set $v(x, \tau) = e^{ax+b\tau} u(x, t)$, and select the $a$ and $b$ values to drop terms.

## 5.5 Exercises

1. Return to Equation 5.4 and collect all terms with a $(\frac{\partial V}{\partial S} - \delta S)$ multiple. From this, show that the objective of minimizing risk by introducing the hedge ratio has the effect of dropping all $\Delta S$ terms from $\Delta_{market} \Pi(S, t)$.
2. Let $f(S, t) = 1/S$. With Itô's Lemma, what is the expression for $\Delta f(S, t)$?
3. A key step in deriving the Black–Scholes Equation was to set the value of $\Delta_{market} \Pi$ equal to what would be gained at a bank. This comparison captures the value of the option relative to interest rates on the market.

   (a) Let $V = C_E$. Instead of setting Equation 5.6 equal to bank rates, compare it to what would be obtained with $\Pi = P_E(S, t) + \beta S$. Find the resulting equation.
   (b) Find the Black–Scholes Equation for $\Pi = C_E(S, t) + P_E(S, t) - \delta S$ with a comparison to the interest rates. Compare answers and explain differences.

4. Suppose $\Delta S = 4S^2 \Delta t + 6S \Delta X$ with the usual assumptions on $\Delta X$. First find the form of Itô's Lemma. Then find the corresponding Black–Scholes Equation.
5. Suppose $\Delta S = 5\Delta t + 6S^2 \Delta X$. First find the form of Itô's Lemma. Then find the corresponding Black–Scholes Equation.
6. Find the boundary conditions for the Black–Scholes Equation for $P_{100}(S, t)$. Remember, this question asks you to find $P_{100}(S, T)$, $P_{100}(0, t)$, and $P_{100}(S, t)$ as $S \to \infty$.
7. Solve the following:

   (a) $y'' + 2y' - 3y = 0$ with $y(0) = 4$, $y'(0) = 0$.
   (b) $2y'' - y' - 3y = 0$ with $y(0) = 6$, $y'(0) = 4$.
   (c) $y'' - 3y' + 2y = 0$ with $y'(0) = 2$, $y'(0) = 0$.
   (d) $xy' + 4y = 0$, $y(1) = 4$.
   (e) $xy' - 3y = 0$, $y(2) = \frac{1}{2}$.
   (f) Solve $x^2 y'' + xy' - y = 0$ where $y(1) = 2$, $y'(1) = 0$.
   (g) $x^2 y'' - 5xy' + 8y = 0$ with $y(1) = 3$, $y'(1) = 8$.

8. Given

$$\frac{\partial V}{\partial \tau} = 2S^2 \frac{\partial^2 V}{\partial S^2} + S \frac{\partial V}{\partial S} - V,$$

   find a change of variable of $S$ to $x(S)$ so that this equation has constant coefficients.
9. Do the same for the Black–Scholes Equation (after using $t = -\tau$)

$$\frac{\partial V}{\partial \tau} = \frac{1}{2} \sigma^2 S^2 \frac{\partial^2 V}{\partial S^2} + rS \frac{\partial V}{\partial S} - rV.$$

# Chapter 6
# Solutions of Black–Scholes

## 6.1 The Heat Equation and $C_E(S, t)$ Solution

Because many introductory courses in partial differential equations solve the heat equation, it is not necessary to do so here. Instead, the emphasis will be to explain the various terms embedded in the solution: All reflect those change of variables that converted the Black–Scholes Equation into the heat equation. These changes must be reinserted into a heat equation solution to transform it into a Black–Scholes solution.

The form of the $C_E(S, t)$ boundary conditions makes it reasonable to expect that the general solution has the form

$$C_E(S, t) = S \times \text{(modifying terms)} - Ee^{-r(T-t)} \times \text{(modifying terms)}, \quad (6.1)$$

where the modifying terms reflect solutions for the heat equation. This is the case. What is needed is a solution for the heat equation.

A surprise (to the uninitiated) is that a normal distribution's PDF, with mean zero and variance $2\tau$,

$$u(x, \tau) = \frac{1}{2\sqrt{\pi \tau}} e^{-\frac{x^2}{4\tau}}, \quad (6.2)$$

satisfies the heat equation Equation 5.11! This assertion can be verified by computing and comparing the indicated partial derivatives. For instance,

$$\frac{\partial u}{\partial \tau} = \frac{1}{2\sqrt{\pi}} \left[ -\frac{1}{2}\tau^{-3/2}e^{-\frac{x^2}{4\tau}} + \tau^{-1/2}\{-\frac{x^2}{4\tau^2}\}e^{-\frac{x^2}{4\tau}} \right].$$

What remains to be computed are the partials with respect to $x$, which are left to the reader.

© Springer Nature Switzerland AG 2019
D. G. Saari, *Mathematics of Finance*, Undergraduate Texts in Mathematics,
https://doi.org/10.1007/978-3-030-25443-8_6

Another form of the solution is

$$u(x, \tau) = \frac{1}{2\sqrt{\pi \tau}} \int_{-\infty}^{\infty} u_0(s, 0) e^{-\frac{(s-x)^2}{2\tau}} \, ds, \tag{6.3}$$

where $u_0(s, 0)$ represents the boundary values. It is here, by using the $C_E(S, T) = \max(S - E, 0)$ boundary condition, that the general form of Equation 6.1 emerges. More precisely, Equation 6.3 makes it reasonable to anticipate that the modifying terms of Equation 6.1 involve integrals resembling $N(x)$, which is the cdf of the normal distribution.

Indeed, the solution for $C_E(S, t)$ is

$$C_E(S, t) = SN(d_1) - E e^{-r(T-t)} N(d_2) \tag{6.4}$$

where

$$d_1 = \frac{\ln(S/E) + (r + \frac{\sigma^2}{2})(T - t)}{\sigma \sqrt{T - t}} \tag{6.5}$$

and

$$d_2 = d_1 - \sigma \sqrt{T - t} = \frac{\ln(S/E) + (r - \frac{\sigma^2}{2})(T - t)}{\sigma \sqrt{T - t}}. \tag{6.6}$$

We have the answer! In fact, using this expression with appropriate values for the parameters, the various prices for a particular Call (as, for instance, $S$ changes) can be computed and printed off to handle the business of a day.

Equations 6.4, 6.5, 6.6 may have an intimidating appearance. To bring this solution down to a comfort zone, what follows is an explanation of what it means and how it adds to our understanding of options.

## 6.2   Source of $C_E(S, t)$ Terms

A way to grasp the meaning of the $C_E(S, t)$ solution (Equations 6.4, 6.5, 6.6) is to dissect its structure. With the

$$C_E(S, t) = S[\text{modifying terms}] - E e^{-r(T-t)} [\text{modifying term}]$$

formulation of a Black–Scholes solution, somewhere—most surely in the modifying components—forms of the heat equation solution should appear. This happens: The modifying terms are given by

$$N(d) = \frac{1}{\sqrt{2\pi}} \int_{-\infty}^{d} e^{-\frac{x^2}{2}}\, dx \qquad (6.7)$$

from the normal distribution *and* a solution of the heat equation. These are the $N(d_1)$ and $N(d_2)$ terms in Equation 6.4.

Now turn to the $d_1$ and $d_2$ expressions, which are dominated by the $\ln(S/E)$ term. The $\ln(S/E)$ value is positive if $S > E$ (so with increasing $S$ values, $d_1$ and $d_2$, along with $N(d_1)$ and $N(d_2)$, increase in value) and it is negative if $S < E$. Rather than a surprise, it is reasonable to expect that $\ln(S/E)$ is part of the solution because of its role (Equation 5.34) in eliminating the variable $S$ and $S^2$ coefficients from the Black–Scholes Equation. Similarly, $r - \frac{\sigma^2}{2}$ is a term that arose when computing the second derivative of this change of independent variable. (See Equation 5.33.)

The denominators in Equations 6.5, 6.6 make sense. According to Theorem 4, this $\sigma\sqrt{T-t}$ value is related to the standard deviation of $\ln(S(t))$, which makes its role as a denominator quite reasonable. Indeed, according to Equation 4.18, $N(d)$ resembles what was used to find the PDF for $S(t)$. Consequently, the modifying terms of Equation 6.4 are related to changes in $S(t)$ values as captured in Section 4.3.

To appreciate the role played by the $d_1$ and $d_2$ numerators, both include

$$\ln\left(\frac{S}{E}\right) + r(T - t) = \ln\left(\frac{S}{Ee^{-r(T-t)}}\right). \qquad (6.8)$$

This means that the numerators of $d_1$ and $d_2$ compare the current price of the asset with the present value of the strike price, which again is an expected relationship. In this manner, the $C_E(S, t)$ value is tied in with the market (captured by the $r$ value) and action of the asset as captured by the standard deviation of $\ln(S)$.

Using Equation 6.8, the $d_j$ expressions become

$$d_1 = \frac{\ln\left(\frac{S}{Ee^{-r(T-t)}}\right)}{\sigma\sqrt{T-t}} + \frac{\sigma}{2}\sqrt{T-t}, \quad d_2 = \frac{\ln\left(\frac{S}{Ee^{-r(T-t)}}\right)}{\sigma\sqrt{T-t}} - \frac{\sigma}{2}\sqrt{T-t}, \qquad (6.9)$$

which catches the $d_1 - d_2 = \sigma\sqrt{T-t}$ relationship.

## 6.3  Interpretation

As true with any newly derived equation, the WGAD concern must be addressed. Beyond providing pricing information for Calls and Puts (which already is valuable), what other information can be extracted from Equations 6.4, 6.9?

A first answer is that the equations indicate how market effects can change a Call's value. For instance, should $t$, the time, be near the expiration date (so $(T - t) \approx 0$), the denominators of $d_1$ and $d_2$ are arbitrarily small. This requires the $d_1$ and $d_2$ values to be heavily determined by the numerator value of $\ln\left(\frac{S}{Ee^{-r(T-t)}}\right)$.

To illustrate, should the spot price be larger than the present value of $E$, which is $S(t) > Ee^{-r(T-t)}$, then $\ln(S/Ee^{-r(T-t)}) > 0$, which (with $(T-t) \approx 0$) forces $d_1$ and $d_2$ to assume arbitrarily large values. In turn, $N(d_1)$ and $N(d_2)$ have values close to unity, which means that $C_E(S,t) \approx S - Ee^{-r(T-t)}$. This makes sense; it asserts that near the expiration date $T$, the value of $C_E(S,t)$ is close to that of the boundary condition for $C_E(S,T)$.

Similarly, should $S(t) < Ee^{-r(T-t)}$, then $d_1$ and $d_2$ have large *negative* values, which force $N(d_1), N(d_2)$ to be close to zero and $C(S,t) \approx 0$. This also makes sense; should $S$ be below the present value of $E$ near the expiration date, there probably is not enough time for prices to change to allow $S \geq E$ at expiration date. So the $C_E(S,t)$ value should be close to the $C_E(S,T)$ boundary value, which is zero.[1]

Equations 6.4, 6.9 include other variables, which permit developing intuition about what can happen with changes in volatility ($\sigma$), interest rates ($r$), strike price ($E$), and various combinations. As an example, suppose the market is quiet, which means that $\sigma$ has a small value. To develop intuition about what this means, consider the extreme case $\sigma \approx 0$.

With a small $\sigma$ value, the $\sigma\sqrt{T-t}$ terms has minimal consequences (Equation 6.9); this means that $d_1$ and $d_2$ are close in value to

$$d_1, d_2 \approx \frac{\ln\left(\frac{S}{Ee^{-r(T-t)}}\right)}{\sigma\sqrt{T-t}}. \tag{6.10}$$

If these small $\sigma$ values are accompanied with $\frac{S}{Ee^{-r(T-t)}} > 1$ ($S$ exceeds the present value of $E$), then $d_1$ and $d_2$ have large positive values. This means that should $\sigma \to 0$, then $d_1 \to \infty$ and $d_2 \to \infty$, which requires $N(d_1), N(d_2) \to 1$. Consequently, with an almost flat market but where $S$ exceeds the present value of $E$, expect the Call's value to satisfy $C_E(S,t) \approx S - Ee^{-r(T-t)}$.

Similarly, should $S$ be smaller than the present value of $E$ (i.e., $S < Ee^{-r(T-t)}$) the Equation 6.10 numerator is negative. If this is accompanied with a very flat market, it follows that $d_1, d_2 \to -\infty$ as $\sigma \to 0$, so $N(d_1), N(d_2) \to 0$. In turn, $C_E(S,T) \approx 0$.

These assertions make sense; if the market is not sufficiently active, there is no reason to expect much change. Consequently, $C_E(S,t)$ reflects either how the current value of $S$ differs from the present value of $E$, or it is near zero reflecting the $C_E(S,T) = \max(S_E, 0)$ constraint.

It is left to the reader to determine what happens when the market has a large $\sigma$; that is, the volatility is high. Here $N(d_1)$ differs from $N(d_2)$, which leads to different interpretations. Other issues involve changes in several variables, such as

---

[1]Recall, if $S < Ee^{-r(T-t)}$, then $S < E$.

if $\sigma \to 0$ when $t \to T$. These double limit concerns are left for the interested reader to explore.

## 6.4 Exercises

1. Instead of using $V - \delta S$ in deriving the Black–Scholes Equation, suppose there is a reason (there are many) to use $\Delta_{bank} = r(V - \delta S)\Delta t + d^* \delta S \Delta t$ where $d^*$ is a fixed constant. Find the new Black–Scholes equation.
   The importance of this problem is that it indicates how to create new equations to handle modified situations.
2. Find the solution of the Black–Scholes Equation for $P_E(S, t)$. (Hint: Put–Call Parity Equation)
3. Suppose the market is wild; it is modeled by $\sigma \to \infty$.

   (a) What is the value of a Call?
   (b) What is the value of a Put?
   (c) Explain both answers in terms of finance.

4. Suppose the modeling allows $\sigma \to 0$; i.e. not much is happening.

   (a) What is the value of a Call?
   (b) What is the value of a Put?
   (c) Explain both answers in terms of finance.

5. Suppose the modeling allows $S \to \infty$;

   (a) What is the value of a Call?
   (b) What is the value of a Put?
   (c) Explain both answers in terms of finance.

6. Suppose the modeling allows $S \to 0$,

   (a) What is the value of a Call?
   (b) What is the value of a Put?
   (c) Explain both answers in terms of finance.

7. Suppose the modeling is that $t \to T$.

   (a) What is the value of a Call?
   (b) What is the value of a Put?
   (c) Explain both answers in terms of finance.

8. Suppose interest rates are increasing enough that it can be modeled with $r \to \infty$.

   (a) What is the value of a Call?
   (b) What is the value of a Put?
   (c) Explain both answers in terms of finance.

9. In deriving the Black–Scholes Equation for a call, we used $Port = C_E(S, t) - \delta S$. In deriving the Black–Scholes Equation, use $C_E(S, t) - \delta C_E(S^*, t)$ where $S^*$ is another asset.
10. Derive the Black–Scholes Equation for $C - \delta S$ for *two* commodities, where another option is to invest money in a bank.
11. Find the delta value to reduce risk for a portfolio $C_E(S, t) - \delta P_E(S, t)$.
12. In the last ten minutes, the value of a call jumped $0.50 while the price of the commodity jumped $1. How should the portfolio be adjusted? (Hint: Find the $\delta$ value.)

# Chapter 7
# Partial Information: The Greeks

## 7.1 The $P_E(S, t)$ Solution

Applying the Equation 6.1 argument to $P_E(S, t)$, along with the boundary condition $P_E(S, T) = \max(E - S, 0)$, makes it reasonable to expect that

$$P_E(S, t) = Ee^{-r(T-t)} \times [\text{modifying terms}] - S \times [\text{modifying terms}].$$

This the case. The actual $P_E(S, t)$ solution follows immediately from our powerful friend the Put–Call Parity Equation.

The approach is as suggested in Exercise 2 of the preceding chapter; the Put–Call Parity Equation requires that

$$\begin{aligned} P_E(S, t) &= C_E(S, t) - S + Ee^{-r(T-t)} \\ &= Ee^{-r(T-t)}[1 - N(d_2)] - S[1 - N(d_1)]. \end{aligned}$$

While this expression suffices for applications, such as the exercises in the preceding chapter, a cleaner equation follows by replacing the $[1 - N(d)]$ terms. To do so, notice that

$$1 - N(d) = \frac{1}{\sqrt{2\pi}} \int_{-\infty}^{\infty} e^{-\frac{x^2}{2}} \, dx - \frac{1}{\sqrt{2\pi}} \int_{-\infty}^{d} e^{-\frac{x^2}{2}} \, dx = \frac{1}{\sqrt{2\pi}} \int_{d}^{\infty} e^{-\frac{x^2}{2}} \, dx.$$

The change of variable $s = -x$ shows that

$$1 - N(d) = \frac{1}{\sqrt{2\pi}} \int_{d}^{\infty} e^{-\frac{x^2}{2}} \, dx = \frac{1}{\sqrt{2\pi}} \int_{-\infty}^{-d} e^{-\frac{s^2}{2}} \, ds = N(-d). \tag{7.1}$$

© Springer Nature Switzerland AG 2019
D. G. Saari, *Mathematics of Finance*, Undergraduate Texts in Mathematics,
https://doi.org/10.1007/978-3-030-25443-8_7

Thus, the solution for the Put assumes the more compact form

$$P_E(S, t) = Ee^{-r(T-t)}N(-d_2) - SN(-d_1). \tag{7.2}$$

## 7.2 Here Come the Greeks!

Knowing "what is the current status" at a given moment (e.g., the values of $C_E(S, t)$, $P_E(S, t)$) is fine, but it sure would be valuable to have a preview of "what will be." A start in this direction was made with the exercises at the end of the preceding chapter. But some of those questions involved extreme, unlikely settings for the variables $S$, $\sigma$, $r$, and $\tau = (T - t)$. With thanks to mathematical tools, *partial* information about "what will be" with option $V$ comes from, well, the *partial* derivatives $\frac{\partial V}{\partial v}$ where $v$ represents one of these variables. A reason this must be the case is that a partial derivative provides slope information, which in turn offers a sense of what will happen.

More precisely, by computing the appropriate partial derivatives of the Equations 6.4 and 7.2 solutions of the Black–Scholes Equation, the sensitivity of an option, how it will change with changes of these variables, emerges. The thrust of this WGAD concern is that, to be successful, there is a need to appreciate how changes in $\sigma$, or $S$, or $r$, or . . . affect $C_E(S, t)$ and $P_E(S, t)$ values. As an example, with an increasing volatility, should a $C_E(S, t)$ be sold now or will it be worth more in the near future? The computations are standard, and, with a little mathematical trickery, lead to fairly simple representations. Of importance is the intuition obtained from these terms.

In the examples, the emphasis is on finding $\frac{\partial C_E(S,t)}{\partial v}$. This is because the corresponding $\frac{\partial P_E(S,t)}{\partial v}$ expression follows immediately from the $P_E(S, t) = C_E(S, t) - S + Ee^{-r(T-t)}$ Put–Call Parity Equation. For instance, the second two terms on the right side, $S$ and $Ee^{-r(T-t)}$, do not include the volatility term $\sigma$, so it follows immediately that $\frac{\partial C_E(S,t)}{\partial \sigma} = \frac{\partial P_E(S,t)}{\partial \sigma}$.

Each partial is represented by a member of the Greek family, which leads to the moniker of "the Greeks." As each $\frac{\partial V(S,t)}{\partial v}$ captures information about how a market change in parameter $v$ affects the value of an option, the Greeks become valued tools in managing portfolios and risk.

### 7.2.1 The Hedge Ratio δ Term

Central to the derivation of the Black–Scholes Equation is the hedge ratio parameter $\delta$. This $\delta$ parameter captures the rate of change of the option with respect to $S$ as given by

$$\delta = \frac{\partial V}{\partial S}. \tag{7.3}$$

Importance of $\delta$ comes from its risk management role with the $V(S, t) - \delta S$ hedge. Recall, this expression is used to balance the hedge mixture between the options and the level of going long or short with the commodity (stock).

To be more precise, according to Taylor's series, should the only change be in the value of $S$, then the corresponding change in $V$ is

$$\Delta V \approx \frac{\partial V}{\partial S} \Delta S = \delta \Delta S. \tag{7.4}$$

(In reading this expression, $\Delta$ is the mathematical change in a variable and $\delta$ is the hedge ratio.) Thus if $\delta = \frac{1}{2}$, a change in $S$ will have half that effect in the change in $V$. Here is a portfolio management question: In this example, the $\delta = \frac{1}{2}$ value is given. If you are managing a portfolio, you will need to compute this $\delta$ value. How? Is there a convenient expression for $\delta$?

To answer this question by carrying out computations with $V(S, t) = C_E(S, t)$, the risk avoidance expression is $\delta_C = \frac{\partial C_E(S,t)}{\partial S}$, where the $\delta$ subscript identifies the option. For $V(S, t) = P_E(S, t)$, it is $\delta_P = \frac{\partial P_E(S,t)}{\partial S}$. These two $\delta$ values share a nice relationship, which, courtesy of Put–Call Parity Equation, is

$$\delta_P = \frac{\partial P_E(S, t)}{\partial S} = \frac{\partial C_E(S, t)}{\partial S} - 1 = \delta_C - 1. \tag{7.5}$$

It remains to find a convenient expression for $\delta_C$. Here, the $C_E(S, t) = SN(d_1) - Ee^{-r(T-t)}N(d_2)$ solution makes it tempting to write

$$\delta_C = \frac{\partial C_E(S, t)}{\partial S} = N(d_1). \tag{7.6}$$

Tempting, but maybe wrong: After all, $N(d_1)$ and $N(d_2)$ also are functions of $S$, so they must be included in the partial derivative computation.

A way to be delivered from temptation is to embrace the product and chain rule (to differentiate the integrals) to obtain

$$\frac{\partial C_E(S, t)}{\partial S} = N(d_1) + S \frac{\partial N(d_1)}{\partial S} - Ee^{-r(T-t)} \frac{\partial N(d_2)}{\partial S}, \tag{7.7}$$

where $\frac{\partial N(d)}{\partial S} = \frac{1}{\sqrt{2\pi}} e^{-\frac{d^2}{2}} \frac{\partial d}{\partial S}$. According to Equations 6.5, 6.6,

$$\frac{\partial d_1}{\partial S} = \frac{\partial d_2}{\partial S} = \frac{1}{\sigma\sqrt{T - t}} \left[ \frac{1}{S/Ee^{-r(T-t)}} \right] \frac{1}{Ee^{-r(T-t)}} = \frac{1}{S\sigma\sqrt{T - t}}.$$

Substituting this information into Equation 7.7 leads to

$$\frac{\partial C_E(S,t)}{\partial S} = N(d_1) + \frac{1}{S\sigma\sqrt{2\pi(T-t)}}\left[Se^{-\frac{d_1^2}{2}} - Ee^{-r(T-t)}e^{-\frac{d_2^2}{2}}\right]. \qquad (7.8)$$

What a mess! That bracket term makes Equation 7.8 clumsy to use. Fortunately, this clutter can be cleaned up with the $a^2-b^2 = (a-b)(a+b)$ algebraic relationship. To obtain this expression, factor $e^{-\frac{d_1^2}{2}}Ee^{-r(T-t)}$ out of the bracket to leave behind

$$\left(\frac{S}{Ee^{-r(T-t)}} - e^{\frac{d_1^2-d_2^2}{2}}\right) = \left(\frac{S}{Ee^{-r(T-t)}} - e^{\frac{(d_1+d_2)(d_1-d_2)}{2}}\right). \qquad (7.9)$$

Further help comes from Equations 6.5, 6.6, 6.9, which establishes that

$$d_1 - d_2 = \sigma\sqrt{T-t}$$
$$d_1 + d_2 = 2\frac{\ln\left(\frac{S}{Ee^{-r(T-t)}}\right)}{\sigma\sqrt{T-t}},$$

or, with a fortuitous cancelation,

$$d_1^2 - d_2^2 = (d_1 - d_2)(d_1 + d_2) = 2\ln\left(\frac{S}{Ee^{-r(T-t)}}\right).$$

Consequently,

$$e^{\frac{(d_1+d_2)(d_1-d_2)}{2}} = e^{\ln\left(\frac{S}{Ee^{-r(T-t)}}\right)} = \frac{S}{Ee^{-r(T-t)}}. \qquad (7.10)$$

This familiar term, comparing the spot price with the present value of $E$, means that the bracket on the right-hand side of Equation 7.9 *vanishes!* This, in turn, forces the ugly Equation 7.8 bracket to disappear. *All that remains is the desired Equation 7.6! It is correct!*

Similarly, the $\delta_P$ value (via the Put–Call Parity Equation) is

$$\delta_P = \frac{\partial P_E(S,t)}{\partial S} = \frac{\partial C_E(S,t)}{\partial S} - 1 = N(d_1) - 1 = -N(-d_1), \qquad (7.11)$$

where the last equality reflects an on-going gift from Equation 7.1. Both expressions are easy to remember by secretly embracing the Equation 7.6 approach, which, while not correct, provides an accurate answer.[1]

It is of importance to interpret Equations 7.6 and 7.11. For $C_E(S,t)$, the hedge ratio $\delta_C$ in $C_E(S,t) - \delta_C S$ has the properties:

---

[1] Similarly, canceling the 6's from $\frac{16}{64}$ leads to the correct answer of $\frac{1}{4}$. Pure coincidence.

- According to Equation 7.6, the $\delta_C$ values range from $0 < \delta_C < 1$. This makes sense; with the negative sign of the $-\delta_C S$ term, a positive $\delta_C$ reflects going short. Going short is a tactic embraced to handle a decline in the price of $S$ while a Call is a bet for an increase in $S$. Betting on both sides creates a hedge.

  The positive value of $\delta_C$ with Equation 7.4 means that an increase in $S$ should increase the value of $C_E(S, t)$; similarly, a drop in $S$ would decrease the value of $C_E(S, t)$.

  > For an intuition break, ignore this mathematical argument and devise an explanation based on the properties of $C_E(S, t)$.

- The hedge essentially disappears should $\delta_C \approx 0$, which requires $d_1 \approx -\infty$. It is interesting to explore how this can arise. One possibility is the $S \to 0$ calamity. A more temperate choice is for $S$ to be smaller than the present value of $E$ and for $\sigma\sqrt{T-t} \approx 0$. That is, a combination of a flat market near expiration date would suffice; as discussed earlier, this scenario forces the ugly $C_E(S, t) \to 0$ setting.

  > For an intuition break, $\delta_C = 0$ is a collapse of the $C_E(S, t) - \delta_C S$ hedge. In the specified situations, explain whether and why this makes sense.

- The other extreme of $\delta_C \approx 1$ requires $d_1 \approx \infty$. An associated circumstance could represent a volatile market (large $\sigma$ value) with an incredibly large $S$ value (so, from the previous chapter, $C_E(S, t) \approx S$). More realistically, it could be a small $\sigma\sqrt{T-t}$ value combined with $S$ larger than the present value of $E$.

  > For an intuition break, why should these situations support a $C_E(S, t) - S$ hedge? If $\delta_C \approx 1$ along with an increasing $S$, what is the expected increase in the value of $C_E(S, t)$? Does this make sense?

For $P_E(S, t)$, the hedge ratio $\delta_P$ from the $P_E(S, t) - \delta_P S$ hedging relationship has opposite properties reflecting the change in sign to $\delta_P = -N(-d_1)$.

- The hedge ratio $\delta_P$ values range from $-1 < \delta_P < 0$, where the negative value requires the hedge, $-\delta_P S$, to go long. This makes sense; $P_E(S, t)$ pays off with decreasing prices, so the hedge must include something giving an improved payoff with increased prices; going long is a choice.
- While the hedge is called off when $\delta_P \approx 0$, the circumstances are opposite of that for a call by requiring $-d_1 \approx -\infty$, or $d_1 \approx \infty$. This behavior could reflect a volatile market (large $\sigma$ value) or a small $\sigma\sqrt{T-t}$ value combined with $S > Ee^{-r(T-t)}$.

  > For an intuition break, what do these conditions mean about the corresponding value of $P_E(S, t)$.

- The other extreme of $\delta_P \approx -1$ requires $-d_1 \approx \infty$, or $d_1 \approx -\infty$. One scenario is for $S$ to be less than the present value of $E$ and $\sigma\sqrt{T-t} \approx 0$. Thus, a flat market near expiration date would suffice.

  > For an intuition break, what do these conditions mean about the associated $P_E(S, t)$ value? If $S$ should increase, what happens to the value of $P_E(S, t)$? What is the accompanying intuition to justify this comment?

## 7.2.2   *Changing δ; the Gamma Greek* Γ

The Section 7.2.1 material is intended to develop intuition about what to expect with changes in the market. The approach is to use a partial derivative to determine how the value of $V$ changes with changes in $S$. Because the $\delta$ value was central to the discussion, it becomes of interest to examine what it takes to change $\delta$ values.

What a can of worms! Answers can crawl into different directions depending on what variables are emphasized: Is the interest in the $\frac{\partial \delta}{\partial \sigma}$ expression, which determines how the volatility of the marker affects the hedge ratio? How about the interest rate $r$, which would call for computing $\frac{\partial \delta}{\partial r}$? All of these terms are of interest, but the choice used here is how $\delta$ changes with changes in $S$. This leads to the "Gamma" Greek value of

$$\Gamma = \frac{\partial \delta}{\partial S} = \frac{\partial}{\partial S} \left[ \frac{\partial V(S,t)}{\partial S} \right] = \frac{\partial^2 V(S,t)}{\partial S^2}. \tag{7.12}$$

According to this expression, $\Gamma$ is the second derivative, or the "acceleration" of changes in $V$. With $V(S,t) = C_E(S,t)$, Equation 7.12 becomes

$$\Gamma = \frac{\partial \delta_C}{\partial S} = \frac{\partial^2 C_E(S,t)}{\partial S^2} = \frac{\partial N(d_1)}{\partial S} = \frac{1}{\sqrt{2\pi}} e^{-\frac{d_1^2}{2}} \frac{\partial d_1}{\partial S}$$
$$= \frac{1}{S\sigma\sqrt{2\pi(T-t)}} e^{-\frac{d_1^2}{2}}. \tag{7.13}$$

Notice what seems to be a conflict of notation. Rather than citing $\Gamma_C = \frac{\partial \delta_C}{\partial S}$, where the $\Gamma$ subscript acknowledges the $\delta_C$ source, there is no Equation 7.13 subscript—the reason, it is not necessary. (Why? Hint: See Equation 7.5.)

And so, $\Delta \delta \approx \Gamma \Delta S$. Further highlighting these terms by returning to the ever useful Taylor series where $S$ is the only changing variable,

$$\Delta C_E(S,t) \approx \frac{\partial C_E(S,t)}{\partial S} \Delta S + \frac{1}{2} \frac{\partial^2 C_E(S,t)}{\partial S^2} [\Delta S]^2 = \delta_C \Delta S + \frac{1}{2} \Gamma [\Delta S]^2.$$

Knowing the $\delta_C$ and $\Gamma$ values leads to sharper $\Delta C_E(S,t)$ estimates.

Immediate observations about $\Gamma$ include:

1. An increasing $S$ value requires $\delta_C$ to increase (because the sign of $\Gamma$ is positive), but its rate of growth (Equation 7.13) flattens out.
2. With high volatility (a large $\sigma$ value), $\Gamma$ tends to be stable.
3. The $e^{-\frac{d_1^2}{2}}$ value is influenced by how $S$ differs from the present value of $E$. And so, with low volatility (small $\sigma$ value) and the stock price near the present value of $E$, expect $\Gamma$ to have a larger impact.

### 7.2.3 The Fake Greek—vega ν

Even a glance at the market exposes a changing volatility. This supports the importance of appreciating how changes in $\sigma$ affect an option's value. In the introductory comments of Section 7.2, it was asked whether the value of a Call will go up or down with a change in $\sigma$; answers follow.

This change is measured by "vega" defined as

$$\nu = \frac{\partial V}{\partial \sigma}. \tag{7.14}$$

There is a slight snag: $\nu$ is the Greek symbol for "nu;" not vega. In fact, *nothing* in the Greek alphabet is called vega; vega is a fake Greek. On the other hand, it might be possible to argue that $\nu$ strongly resembles the shape of the nonexistent vega, so that is what it will be called!

It remains to determine how the value of an option changes with changes in volatility. Computing with $V(S, t) = C_E(S, t)$, it follows that

$$\nu = \frac{\partial C_E(S, t)}{\partial \sigma} = S \frac{\partial N(d_1)}{\partial \sigma} - E e^{-r(T-t)} \frac{\partial N(d_2)}{\partial \sigma},$$

where

$$\frac{\partial N(d)}{\partial \sigma} = \frac{1}{\sqrt{2\pi}} e^{-\frac{d^2}{2}} \frac{\partial d}{\partial \sigma}$$

and

$$\frac{\partial d_1}{\partial \sigma} = -\frac{\ln\left(\frac{S}{Ee^{-r(T-t)}}\right)}{\sigma^2 \sqrt{T-t}} + \frac{\sqrt{T-t}}{2}, \qquad \frac{\partial d_2}{\partial \sigma} = -\frac{\ln\left(\frac{S}{Ee^{-r(T-t)}}\right)}{\sigma^2 \sqrt{T-t}} - \frac{\sqrt{T-t}}{2}.$$

Another mess! The value of vega is

$$\nu = \frac{\partial C_E(S,t)}{\partial \sigma} = -\frac{C}{\sigma^2}\left[\ln\left(\frac{S}{Ee^{-r(T-t)}}\right)\right]\{Se^{-\frac{d_1^2}{2}} - Ee^{-r(T-t)}e^{-\frac{d_2^2}{2}}\}$$

$$+ \sqrt{\frac{T-t}{2\pi}}\left[\frac{Se^{-\frac{d_1^2}{2}} + Ee^{-r(T-t)}e^{-\frac{d_2^2}{2}}}{2}\right] \tag{7.15}$$

with positive constant $C = \frac{1}{\sqrt{2\pi(T-t)}}$. Significant help in cleaning up this jumble of terms comes from the derivation of $\delta_C = N(d_1)$; recall, there it was shown that the bracketed term of Equation 7.8 equals zero. The same term appears in Equation 7.15, so

$$v = \sqrt{\frac{T-t}{2\pi}} \left[ \frac{Se^{-\frac{d_1^2}{2}} + Ee^{-r(T-t)}e^{-\frac{d_2^2}{2}}}{2} \right].$$

Using the same approach to obtain the $a^2 - b^2 = (a-b)(a+b)$ exponent and then the Equation 7.10 expression,

$$\begin{aligned} v &= \sqrt{\frac{T-t}{2\pi}} Ee^{-r(T-t)}e^{-\frac{d_1^2}{2}} \left[ \frac{S}{2Ee^{-r(T-t)}} + \frac{e^{\frac{d_1^2-d_2^2}{2}}}{2} \right] \\ &= \sqrt{\frac{T-t}{2\pi}} Ee^{-r(T-t)}e^{-\frac{d_1^2}{2}} \left[ \frac{S}{2Ee^{-r(T-t)}} + \frac{S}{2Ee^{-r(T-t)}} \right] \\ &= S\sqrt{\frac{T-t}{2\pi}}e^{-\frac{d_1^2}{2}}, \end{aligned}$$

(7.16)

with a sharply simpler expression!

The significance of this fake Greek for finance makes it worth highlighting the expression by repeating it.

$$v = \frac{\partial C_E(S,t)}{\partial \sigma} = S\sqrt{\frac{T-t}{2\pi}}e^{-\frac{d_1^2}{2}}.$$

(7.17)

Even though vega captures how $C_E(S,t)$ (and $P_E(S,t)$) vary with volatility, the explicit role played by $\sigma$ is hidden in the $d_1^2$ term. Yet, with an analysis similar to the above, the manner in which vega changes with $S$ and $\sigma$ values can be extracted. Anyway, should interest be restricted to $\sigma$ changes,

$$\Delta C_E(S,t) = v\Delta\sigma.$$

(7.18)

A sample of lessons learned from Equation 7.18 include:

1. An increase in the volatility causes the option price to increase.
2. The amount of increase depends on the magnitude of $v$. As an example, the longer away it is to expiration date (i.e., the larger the $T-t$ value), the larger the increase in the option's price. Similarly, $v$ depends on $S$, so larger $S$ values lead to larger $v$ values.

### 7.2.4  More Members of the Greek Party

The approach is clear: To determine how an option's value changes (locally) with market changes, take its partial derivatives with respect to the relevant variable. The derivations involve standard terms, which ensures that the computations are elementary but occasionally messy. The value added comes from analyzing the expressions to extract what they explain about changes in the value of options.

Some of the other Greeks follow:

- As it must be expected, as expiration date draws nearer, opportunities for an option to change in a desired direction fade away. This behavior is captured with the Greek Theta, $\Theta$, which measures changes in an option as time to the expiration date, $\tau = T - t$, decreases. It is

$$\Theta = \frac{\partial V}{\partial t} = -\frac{\partial V}{\partial \tau}. \tag{7.19}$$

Thus, $\Theta$ measures how fast an option loses value as expiration date approaches. The units are usually in terms of days.

- The derivation of the Black–Scholes Equation compares advantages of a given stock with other opportunities. As this argument revolves about the interest rate, effects of $r$ are to be watched. The Greeks assign this responsibility to Rho, $\rho$, which measures how an option's value changes with changes in the interest rate:

$$\rho = \frac{\partial V}{\partial r}. \tag{7.20}$$

- Vanna is another fake Greek; it captures settings that "vanna" be Greek by measuring how $\delta$ changes with volatility, perhaps with hopes of identifying a wheel of fortune. Thus, for $V$,

$$\text{Vanna} = \frac{\partial \delta}{\partial \sigma} = \frac{\partial}{\partial \sigma} \left[ \frac{\partial V(S,t)}{\partial S} \right] = \frac{\partial^2 V(S,t)}{\partial \sigma \partial S} \tag{7.21}$$

In other words, while Vanna is computed as a partial derivative of $\delta$, it is the mixed partial $\frac{\partial^2 V(S,t)}{\partial \sigma \partial S}$.

- Reaching beyond Vanna, it is of interest to learn how $\delta$ changes with respect to other variables.

It is very easy to go on and on. How does $\rho$ vary with price changes? How does Vanna change with the interest rate? (This leads to a third order derivative of $V$.) What to explore depends on what you need to accomplish.

As an illustration, suppose both $S$ and $\sigma$ are varying, where, for purposes of forecasting, you need a more refined estimate of $\Delta C_E(S,t)$ than what follows from the above. The natural approach is to appeal to the power of Taylor series to have

$$\Delta C_E \approx \frac{\partial C_E}{\partial S} \Delta S + \frac{\partial C_E}{\partial \sigma} \Delta \sigma + \frac{1}{2} \frac{\partial^2 C_E}{\partial S^2} (\Delta S)^2 + \frac{\partial^2 C_E}{\partial S \partial \sigma} \Delta S \Delta \sigma + \frac{1}{2} \frac{\partial^2 C_E}{\partial \sigma^2} (\Delta \sigma)^2.$$

Thanks to the Greeks, this becomes

$$\Delta C_E \approx \delta_C \Delta S + v \Delta \sigma + \frac{1}{2} \Gamma (\Delta S)^2 + Vanna \Delta S \Delta \sigma + \frac{1}{2} \mathcal{TBD} (\Delta \sigma)^2,$$

where $TBD$ (To Be Determined) equals $\frac{\partial^2 C_E}{\partial \sigma^2}$. Because $\frac{\partial^2 C_E}{\partial \sigma^2} = \frac{\partial}{\partial \sigma}\left[\frac{\partial C_E}{\partial \sigma}\right] = \frac{\partial v}{\partial \sigma}$, the answer is $TBD = \frac{\partial v}{\partial \sigma}$, which is not difficult to compute.

Whatever the need, the tools for achieving information are as described here: Take the appropriate partial derivatives.

## 7.3 Exercises

1. The first exercise is to find the various partials of $C_E(S, t)$ in order to obtain a better sense of how changes in variable affect its value.

   (a) Find $\frac{\partial C_E(S,t)}{\partial r}$. What does this partial mean should $r$ increase? Should $r$ decrease?
   (b) Find $\frac{\partial C_E(S,t)}{\partial \sigma}$. What does this partial mean should the value of $\sigma$ increase? (The market is more volatile.) Should $\sigma$ decrease? (The market calms down.)
   (c) Find $\frac{\partial C_E(S,t)}{\partial t}$.

2. This exercise does the same, but now for $P_E(S, t)$.

   (a) Find $\frac{\partial P_E(S,t)}{\partial S}$. What does this partial mean if $S$ increases in value? If it decreases in value?
   (b) Find $\frac{\partial P_E(S,t)}{\partial r}$. What does this partial mean should $r$ increase? Should $r$ decrease?
   (c) Find $\frac{\partial P_E(S,t)}{\partial \sigma}$. What does this partial mean should the value of $\sigma$ increase? (The market is more volatile.) Should $\sigma$ decrease? (The market calms down.)
   (d) Find $\frac{\partial P_E(S,t)}{\partial t}$.

3. Rather than a direct computation, a quick way to find the partials of $P_E(S, t)$ is to use the workhorse of the Put–Call Parity Equation. Using this approach, find the relationships between all of the described partial derivatives of $C_E(S, t)$ and $P_E(S, t)$.

4. Suppose $\delta_P$ has a value close to $-1$. What can be said about the $\delta_C$ value?

5. In Equation 7.15, vega is computed in terms of $C_E(S, t)$. What would be the value had $P_E(S, t)$ been used?

6. Let $\Gamma_C$ equal $\frac{\partial \delta_C}{\partial S}$. What would be the $\Gamma_P$ value? How would Equation 7.5 help in this computation?

7. Suppose $\delta_C = 0.6$ and $\Gamma = 2$, where $\Delta S = \frac{1}{2}$. Find an estimate for $\Delta C_E(S, t)$.

8. Assume that the only two variables of interest are $S$ and $\sigma$. Write down the second order Taylor series approximation for $\Delta C_E(S, t)$. For the various partials, substitute the appropriate Greek name. One term needs a name and a computation, what is it and what is its value? (The purpose of this exercise is to answer the comments of the concluding paragraphs of the chapter.)

9. For a fixed $S = 100$ value, sketch the graph of $\delta_C$ for a small value of $T - t$, and then a large value of $T - t$ as the strike price $E$ varies from 50 to 150.

# Chapter 8
# Sketching and the American Options

Although $C_E(S, t)$ and $P_E(S, t)$ have been analyzed and described in various ways, something is missing. Similar to where a description of the moon shining over a lake nested in the snow-capped mountains is a poor substitute for an actual picture, what is needed is a portrait of these options. And so, we now give an outline how to sketch the graphs. Of added value, the method can be adopted to sketch our new acquaintances, the Greeks, to better understand where they exercise power.[1]

As one might hope, the graphs lead to surprises; surprises that help to understand a new class of options. But first, the pictures.

## 8.1 Using $\delta_C$ and $\delta_P$ to Sketch $C_E(S, t)$ and $P_E(S, t)$

Assistance in drawing the $y = V(S, t)$ curves comes from the curve's slope as given by $\delta = \frac{\partial V(S,t)}{\partial S}$. In this way, the earlier discussion of $\delta$'s properties helps to describe the curves of $y = C_E(S, t)$ and $y = P_E(S, t)$. The approach follows standard calculus lessons: Find appropriate asymptotes.

### 8.1.1 Sketching $C_E(S, t)$

To sketch $y = C_E(S, t)$, start with the reference curve $C_E(S, T) = \max(S - E, 0)$ that describes $C_E$'s behavior at expiration date. In Figure 8.1a, this curve lies on the $S$-axis up to $S = E$, and then it moves off to infinity on the solid line with

---

[1]With a blink of an eye, sketching programs yield these drawings. But, in order to appreciate properties and opportunities of options, we need to go beyond the graphs to understand *why* they possess certain features. This is achieved though finding how to carry out the sketchs.

© Springer Nature Switzerland AG 2019
D. G. Saari, *Mathematics of Finance*, Undergraduate Texts in Mathematics,
https://doi.org/10.1007/978-3-030-25443-8_8

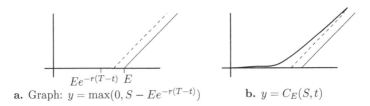

**a.** Graph: $y = \max(0, S - Ee^{-r(T-t)})$          **b.** $y = C_E(S,t)$

**Fig. 8.1** Sketching $C_E(S,t)$

slope of unity. The main difference between this line and second reference curve $y = \max(S - Ee^{-r(T-t)}, 0)$ is that the $S$-axis intercept for the second curve is at the smaller $Ee^{-r(T-t)}$ value. This second curve is on the $S$-axis up to this intercept, and then it becomes the dashed line with slope unity.

Following lessons learned in calculus, key aspects of the $y = C_E(S,t)$ curve come from its asymptotic properties as $S \to 0$ and $S \to \infty$. For the first, earlier computations show that if $S \to 0$, then $C_E(S,t) \to 0$ and $d_1, d_2 \to -\infty$. The $d_1$ values require $\delta_C = N(d_1) \to 0$, so as $S \to 0$ (which requires $C_E(S,t) \to 0$), its slope flattens to approach zero.

Similarly, as $S \to \infty$, then $d_1, d_2 \to \infty$, which force $N(d_1), N(d_2) \to 1$. According to Equations 6.5, 6.6, $C_E(S,t)$ approaches the dashed reference line $y = S - Ee^{-r(T-t)}$, but it never quite hits it. (Remember, $\delta_C < 1$.) The closer $C_E(S,t)$ approaches the dashed line, the closer the slope $\delta_C = N(d_1)$ is to unity.

Between the two extremes, $\delta_C > 0$, which prohibits the curve from experiencing any dips or swanning movements. It remains to find an intermediate point; a natural choice is where the spot price hits the present value of $E$, or $S = Ee^{-r(T-t)}$. A reason for this choice is that it is conveniently located right where the dashed line hits the $S$ axis. A second reason is mathematical ease; with this choice $\ln(S/Ee^{-r(T-t)}) = 0$, so Equation 6.9 becomes

$$d_1^* = \frac{\sigma}{2}\sqrt{T-t}, \quad d_2^* = -\frac{\sigma}{2}\sqrt{T-t}, \tag{8.1}$$

where $N(d_1^*) = N(\frac{\sigma}{2}\sqrt{T-t}) > \frac{1}{2}$ and because $-d_1^* = d_2^*$, $N(d_2^*) = 1 - N(d_1^*) < \frac{1}{2}$. Consequently, the height of the curve at this point is

$$C_E(Ee^{-r(T-t)}, t) = Ee^{-r(T-t)}(N(d_1^*) - N(d_2^*)) = Ee^{-r(T-t)}(2N(d_1^*) - 1) \tag{8.2}$$

with slope $\delta_C = N(d_1^*)$.

The height of the curve and size of the slope at this intermediate point depend on the $\frac{\sigma}{2}\sqrt{T-t}$ value; smaller values (e.g., near expiration date and/or a flat market) place $N(d_1) \approx \frac{1}{2}$, while larger values (e.g., a volatile market) force a larger $N(d_1)$ value. With this information, the curve $y = C_E(S,t)$ can be sketched, as in Figure 8.1b.

### *8.1.2 Sketching $P_E(S, t)$*

A similar approach holds to sketch

$$y = P_E(S, t) = Ee^{-r(T-t)}N(-d_2) - SN(-d_1).$$

Namely, find the two reference lines, the two asymptotic behaviors, an intermediate point, and then draw a curve connecting all elements.

The first reference line describes what happens at expiration date, where $P_E(S, T) = \max(E - S, 0)$. In Figure 8.2a, this is the positive part of solid line $y = E - S$ with intercept at $S = E$ and slope $-1$. The second reference line, depicted by the Figure 8.2a dashed line, is the $y = \max(Ee^{-r(T-t)} - S, 0)$ curve. As true with Figure 8.1a, the intercept of the dashed line on the $S$-axis is $Ee^{-r(T-t)}$, which is *smaller* than $E$. As the line's slope is $-1$, the dashed line meets the $y$-axis at $y = Ee^{-r(T-t)}$.

For the first of the two asymptotes, as $S \to \infty$, $d_1, d_2 \to \infty$, which forces $P_E(S, t) \to 0$. The slope of the curve is $\delta = -N(-d_1)$, which approaches zero for large $S$ values. So, as $S \to \infty$, the $y = P_E(S, t)$ curve flattens out while approaching the $S$-axis.

The other direction has $S \to 0$, so $d_1, d_2 \to -\infty$, or $-d_1, -d_2 \to \infty$ and $N(-d_1), N(-d_2) \to 1$. Consequently,

$$\text{as } S \to 0, \ P_E(S, t) \to Ee^{-r(T-t)}.$$

The left limit of the curve is the point $(0, Ee^{-r(T-t)})$, which is where the dashed line meets the $y$-axis. Away from the limit point, the slope (given by $\delta = -N(-d_1)$) satisfies $\delta > -1$, so, although the $y = P_E(S, t)$ curve approaches the $y = \max(Ee^{-r(T-t)} - S)$ reference line as $S \to 0$, it is always a bit to the right; it never touches the dashed boundary.

For the same reasons $S = Ee^{-r(T-t)}$ was used as an intermediate point in Figure 8.1b, it used here—with an added bonus that the $d_1^*, d_2^*$ values already are computed (Equation 8.1)! Thus the height of the curve at this point is $P_E(Ee^{-r(T-t)}, t) = Ee^{-r(T-t)}\left[N(-d_2^*) - N(-d_1^*)\right]$. But $-d_1^* = d_2^*$, so

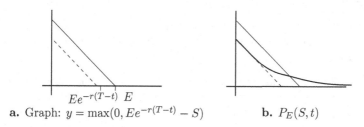

**a.** Graph: $y = \max(0, Ee^{-r(T-t)} - S)$      **b.** $P_E(S, t)$

**Fig. 8.2** Sketching $P_E(S, t)$

$$P_E(Ee^{-r(T-t)}, t) = Ee^{-r(T-t)} \left[ N(d_1^*) - N(d_2^*) \right]$$

has the same $Ee^{-r(T-t)}(2N(d_1^*) - 1)$ value of Equation 8.2. The slope differs; according to the Put–Call Parity Equation, it is $\frac{\partial P_E(Ee^{-r(T-t)}, t)}{\partial S} = -N(-d_1^*) =$ $\frac{\partial C_E(Ee^{-r(T-t)}, t)}{\partial S} - 1 = N(d_1^*) - 1$. This information leads to the Figure 8.2b sketch.

### 8.1.3  Comparing Curves

Some observations from the above computations:

1. The structure of the two curves, where $y_C = C_E(S, t)$ moves from zero on the $S$ axis to infinity while $y_P = P_E(S, t)$ moves from $(0, Ee^{-r(T-t)})$ to the $S$ axis ensures that they intersect in a unique point. As computed above, the curves $C_E(S, t)$ and $P_E(S, t)$ intersect precisely where $S = Ee^{-r(T-t)}$; it is where $S$ equals the present value of $E$.
2. The Figure 8.1b sketch of $y_C = C_E(S, t)$ proves that the curve remains separated from the reference line $y = \max(S - E, 0)$.
3. In contrast, if $t < T$, then the curve $y_P = P_E(S, t)$ must cross the $y = \max(E - S, 0)$ line. As shown in the next section, this difference makes a difference.
4. According to the Put–Call Parity Equation, at each $S$ value, the slopes of the two curves are related in that they satisfy $\frac{\partial P_E(S, t)}{\partial S} = \frac{\partial C_E(S, t)}{\partial S} - 1$. This expression reflects the useful equality $-N(-d) = N(d) - 1$.

## 8.2  Arbitrage and the American Option

Here is a question that always should be asked: Are there any profit opportunities hiding in these curves? To explore this issue, let's jazz up the discussion by adding flexibility in the use of an option. For an intuition break, what should this mean concerning the value of an option, such as a Put or Call?

It is reasonable to expect that an added feature would be considered only if it provides an advantage. In turn, this suggests:

1. An added feature is advantageous if the option gains value.
2. A way to identify added value is to determine whether the shift in rules provides arbitrage opportunities.

To be specific, consider the *American option;* this is where the option can be exercised at any time prior, or equal to the expiration date. There are stipulations; it would be uncivilized to awaken someone at 3:25 am on a Sunday morning, when a person may be recovering from adult beverages, to exercise a Put within the next five minutes. But even with reasonable restrictions, the American Put appears to have more flexibility than the described properties of a European Put.

Does it? Is an American Put really more valuable? The added advantages sound fabulous, but could they be fanciful, such as a guarantee for doubled earnings should a unicorn stumble into your home? What needs to be determined is whether, when, and why the new rules provide advantage. To sharpen the arbitrage sensor, check Figure 8.2b to determine whether any opportunities are waiting to be exploited if a European Put is suddenly converted into an American Put.

### 8.2.1 Simple Geometry for Puts

Before answering the posed question, consider some elementary geometry associated with a line with slope $-1$ as depicted in Figure 8.3a. Of interest is that isosceles triangle with sides of length $p_1$ and its hypotenuse with slope $-1$ that hits the $x$ axis at $x = e_1$. If $s_1$ is the distance from the origin to the right-angle vertex of the triangle, then, trivially as displayed on the $x$-axis, $s_1 + p_1 = e_1$.

Taking advantage of the triangle, another way to express the $s_1 + p_1 = e_1$ equality is to traverse the $x$-axis to $x = s_1$, and then take a sharp left turn to move $p_1$ units in the $y$ direction to hit the slanted line. This line has the expression $y = e_1 - x$, so at $x = s_1$, it is $p_1 = e_1 - s_1$, which is the earlier $e_1 = s_1 + p_1$. Another obvious comment is that replacing $p_1$ with $p_1^* < p_1$ (so in the $y$ direction, $p_1^*$ is lower than $p_1$) means that $s_1 + p_1^* < e_1$.

A strong clue that these comments have something to do with finance is that, well, they are included in a book on finance. To see where this fits in, notice from Figure 8.3b that the cost of a European Put must eventually fall *below* the $y = \max(E - S, 0)$ line! To detect possible actions, consider a stock where its current price is $S_1$, as in Figure 8.3b. At this price, the value of the Put is $P_E(S_1, t)$, which is *below* the $y = E - S$ curve. For precisely the same reason $s_1 + p_1^* < e_1$, it follows that

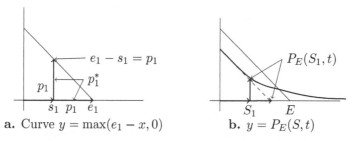

**a.** Curve $y = \max(e_1 - x, 0)$    **b.** $y = P_E(S, t)$

**Fig. 8.3** American Put

$$S_1 + P_E(S_1, t) < E. \tag{8.3}$$

Any inequality that crops up in finance must be examined to determine whether it has some arbitrage grease. Remember, key to arbitrage is

*buy low, sell high!*

The "sell high" phrase calls attention to the role of a Put; it permits *selling* the object for the specified strike price of $E.

To check whether there are "buy low" opportunities, in order to exercise the American Put, two items are needed:

1. a Put, $P_E(S, t)$, to exercise and
2. the stock to sell.

To assemble everything requires *buying* the Put for $\$P_E(S_1, t)$ and *buying* the stock for $\$S_1$. The sum of expenditures is on the left-hand side of Equation 8.3. Because the "buy low" cost is lower than the $E rewards, *immediately* exercise the Put, which means selling the stock for $E. The Equation 8.3 difference of $E - [S_1 + P_E(S_1, t)]$ is the arbitrage profit! Stated more precisely,

whenever the cost of an American $P_E(S, t)$ falls below the line $y = E - S$, *arbitrage opportunities exist!*

To illustrate with numbers, suppose Katrina has an American Put with the value $P_{100}(60, t) = 30$. This means that the strike price is $100, the current value of the asset is $S = \$60$, and the cost of the Put is $30. The cost of buying a Put and the stock, which is the left-hand side of Equation 8.3, is $30 + 60 = 90$. The strike price, which is the right-hand side of Equation 8.3, is 100. Because $90 < 100$, Katrina has a "buy low, sell high" opportunity, which means that

- Katrina should buy $P_{100}(60, t)$ for $30.
- She should buy the asset for $60. So far, the Put and the asset have been acquired for a total cost of $90.
- Exercise the Put; that is, Katrina should sell the asset for the strike price of $100.
- She now can enjoy the guaranteed arbitrage profit of $100 - \$90 = \$10$.

Intuition break. Whenever the cost of an American $P_E(S, t)$ falls below the line $y = E - S$, *arbitrage opportunities exist!* Are there any opportunities should $P_E(S, t)$ be above, or on the line, $y = E - S$? A Figure 8.3a type analysis, or a "buy high, sell low" story, will give the answer.

## 8.2.2   Arbitrage with a Call

What about a Call? To sell-high, buy-low, the first step is to determine what needs to be bought and what can be sold. To do anything, the American Call can be purchased at the cost of $\$C_E(S, t)$. Exercising the Call requires buying the stock at the cost of $E.

These are the purchases: Something needs to be sold, and it is the newly purchased commodity with the current price of \$S. If there is a buy-low, sell-high opportunity, the arbitrage maxim is expressed as

$$C_E(S, t) + E < S,$$

which is the same as

$$C_E(S, t) < S - E \qquad (8.4)$$

For arbitrage opportunities, Equation 8.4 requires the cost $\$C_E(S, t)$ of the Call to be *below* the $y = S - E$ line.

Describing this story in a different manner, convert the Figure 8.3a geometry into a Figure 8.4a setting with the line $y = \max(x - e_2, 0)$. In this figure, $e_2$ is the first segment on the $x$-axis, while $c_2$ is the segment from $e_2$ to $s_2$. Again, the construction shows that $e_2 + c_2 = s_2$.

To translate this equality into the properties of the $y = x - e_2$ line, at $s_2$ take a sharp left turn and move $c_2$ units upwards. This construction hits the $y = x - e_2$ at $c_2 = s_2 - e_2$, which is the same $s_2 = c_2 + e_2$ expression. Replacing $c_2$ with the smaller $c_2^* < c_2$, which is below the $y = x - e_2$ line at $x = s_2$, shows it is insufficient: $e_2 + c_2^* < s_2$. The "$s_2 - (e_2 + c_2^*)$" difference is illustrated with the Figure 8.4a gap on the $x$-axis between $e_2$ and $c_2^*$.

Transferring this discussion to Calls, if the graph of an American $C_E(S, t)$ ever dips below the $y = S - E$ line, then, as illustrated in Figure 8.4b (and computed with Equation 8.4), the Figure 8.4a geometry applies.

Here is a problem: As computed in Figure 8.2b, it is *impossible* (should $t < T$) for a Call to even meet the $y = S - E$ line, so it cannot dip below it. Thus, this discussion has the flavor of "Here come the unicorns!"–fanciful but not of real value. Actually, this is *not* the case. What provides value to the argument is that, as shown in the next chapter, by including embellishments such as dividends and other features, it is possible for the inequality of Equation 8.4 to arise and allow arbitrage!

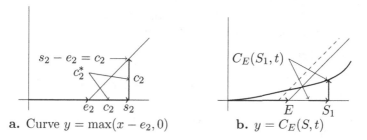

a.  Curve $y = \max(x - e_2, 0)$          b.  $y = C_E(S, t)$

**Fig. 8.4** American Call

As an illustration with numbers, suppose $C_{60}(80, t) = 10$ where the strike price is $60, the current value of the asset is $80, and the current price of the Call is $10. This issue is to determine whether this "American option" setting, where the Call can be exercised at any time, provides personal opportunities.

The "buy low, sell high" arbitrage expression is as follows:

- *Buy $C_{60}(80, t)$ for $10.*
- Immediately exercise the contract in order to *buy* the asset for the strike price of $60. So far, the total cost of purchases—the Call and the asset—is $70.
- *Immediately* sell the asset for the current price of $80.
- Enjoy the guaranteed arbitrage profit of $80 − $70 = $10.

## How to Lose Money

To appreciate the general behavior of an American option, suppose $P_{100}(60, t) = \$45$. To explore whether there are any arbitrage opportunities, consider what happens should the above approach be mimicked:

- Buy the option for $45.
- Buy the asset for $60; the asset and Put have been purchased for $105.
- Exercise the Put to sell asset at the strike price of $100.
- Suffer the guaranteed *loss!!* of $100 - $105 = -$5.

The problem, of course, is that

$$P_{100}(60, t) = 45 > E - S = 100 - 60.$$

More generally, if $P_E(S, t)$ is above or on the line $y = E - S$, forget trying to find arbitrage openings.

Similarly, if $C_{60}(80, t) = 25$, a failed attempt at arbitrage would be to:

- Buy $C_{60}(80, t)$ for $25.
- Immediately exercise the contract to buy the asset for the strike price of $60. So far, the Call and the asset have been acquired for a total cost of $85.
- Immediately sell the asset for the current price of $80.
- Suffer the guaranteed loss of $80 − $85 = −$5.

The problem is that

$$C_{60}(80, t) = 25 > S - E = 80 - 60 = 20.$$

Consequently, if $C_E(S, t)$ is above or on the line $y = S - E$, the arbitrage opportunities have dried up.

### 8.2.3 New Rules; the American Option

This American option differs from the European option by introducing flexibility in when the option is exercised; it need not be at $t = T$. By providing added opportunities, it is reasonable to expect that the value, the cost, of such an American option is greater than that of a European one. This is the case.

The increase in value follows from the arbitrage examples: With the Call example, a reasonable individual would *not* price $C_{60}(80, t) < 80 - 60 = 20$; doing so is an invitation to be exploited by providing an arbitrage opportunity. Arbitrage, then, is the market pressure that can be expected to increase the price to at least $C_{60}(80, t) = 20$. More generally, for an American Call,

$$C_E(S, t) \geq S - E. \tag{8.5}$$

A similar argument holds for the Put arbitrage example. To assign the price of $P_{100}(60, t) < 100 - 60 = 40$ is to welcome investors to exploit you. This means that the price of the American $P_{100}(60, t)$ is at least $\$100 - \$60 = \$40$. More generally, for an American Put,

$$P_E(S, t) \geq E - S. \tag{8.6}$$

The emphasis of above two examples is that the market pressure of arbitrage is *not* available if $P_E(S, t) \geq E - S$ for Puts, or if $C_E(S, t) \leq S - E$ for Calls. In these settings, expect the usual market pressures to apply.

To sketch the American Put, modify the sketch for the European Put in the following way:

- If the sketch for the European Put is above the $E - S$ line, that is the sketch for the American Put. This reflects the fact that in this region, both the American and the European Puts are subject to market pressures as reflected by the Black–Scholes Equation.
- Once the sketch for the European Put hits the $E - S$ line and seeks to go below, alter the the American Put sketch to slide along the $E - S$ line as $S$ decreases in value.
- This sketch has a sharp corner caused by bending the sketch of the European Put to be on the $E - S$ line. In fact, it has a smooth bend, which again is caused by arbitrage (here the anticipation of the above scenarios play a role). All of this can be made mathematically precise by modeling the American option with variational equations, but, while interesting, this extends beyond the scope of these notes.

A similar description holds for the sketch of the American Call; that is left for the reader.

## 8.3  Exercises

1. This exercise involves a straddle with European options $\Pi = P_E(S, t) + C_E(S, t)$ for $t < T$.

   (a) Find the $S$ value that is a critical point of $\Pi$.
   (b) Show that the slop of $\Pi$ is increasing for $S$ larger than this critical point, and the slope is decreasing for $S$ smaller than this critical point. That is, the critical point is a minimum.
   (c) Sketch $\Pi$.
   (d) Suppose these are American options; find the sketch of $\Pi$.

2. When sketching $y = C_E(S, t)$ and $y = P_E(S, t)$, it was discovered that $C_E$ and $P_E$ have the same value at $S = E^{-r(T-t)}$. Derive this result by using the Put–Call Pariety Equation.

3. Sketch both the European options $C_{50}(S, t)$ and $P_{50}(S, t)$ for $t < T$. Explain the graph.

4. With European options, sketch the strangle $P_{90}(S, t) + C_{110}(S, t)$.

5. Suppose the European $P_{40}(20) = 10$ suddenly becomes an American option. Are there arbitrage opportunities? Explain by carrying out computations.

6. Suppose the European $P_{40}(20) = 30$ suddenly becomes an American option. Are there arbitrage opportunities? Give an explanation.

7. Suppose the European $C_{80}(100) = 10$ suddenly becomes an American option. Are there arbitrage opportunities? Give an answer in terms of money earned or lost.

8. Suppose the European $C_{80}(100) = 30$ suddenly becomes an American option. Are there arbitrage opportunities? Again, explain in terms of money lost or gained.

9. Sketch both the American options $C_{50}(S, t)$ and $P_{50}(S, t)$ for $t < T$. Explain what is happening with each graph.

10. Should all variables other than $S$ remain fixed, sketch $\delta_C$ and $\delta_P$. In doing so and as with the sketches of $C_E(S, t)$, first compute the asymptotic behavior as $S \to 0$ and as $S \to \infty$. Then use the sign of $\Gamma$ as given in Equation 7.13. Also notice from Equation 7.13 how the value of $\Gamma$ (or the $d_1$ value near the present value of $E$) allows a more rapid jump near $S = E$ as $t \to T$. What does this mean about the graph?

11. Important to the market is the effects of volatility. Find a crude sketch of $v$ (Equation 7.12) with respect to $S$.

# Chapter 9
# Embellishments

A delight of this topic is that it is possible to go on and on and on. But closure is needed somewhere, and it is with this chapter. The farewell message is to stress that the powerful tools derived in this book can be used elsewhere: This concluding chapter suggests how and where. At this stage, for instance, the reader probably can develop at least a partial explanation for other topics encountered on the market. It may take some imagination to find a surrogate for inflation, or pollution, or ..., but being able to do so is what offers a personal advantage.

Who knows, the reader might encounter an opportunity to invest in an option involving the weather (they exist; for instance [4, 7]!), or pollution, or ... The problem is to understand how to hedge and where arbitrage opportunities arise. In earlier chapters, a crucial form of hedging involved the form $C_E(S, t) - \delta S$, where the Call was balanced by going short on the asset. With pollution, or weather, or ... it is difficult to go short on these objects. Thus, a surrogate is needed.

## 9.1 Bonds

To be more precise, consider bonds. But first, recall that a bond is a debt obligation. A local school may want to build a swimming school; a corporation may need a new factory. To support these expensive projects, the amount of borrowed money may exceed what can be obtained from a bank. So, to secure appropriate amounts of money, the organization issues a bond. A person who buys a bond is loaning money to the organization—bonds represent debt to the organization. At a "maturity date" $T$, the amount borrowed—the face value—must be paid off.

Similar to how money in the bank receives interest, or mortgage payments on a new home include interest charges, the organization, in repaying the bond, makes interest payments on a regular basis. One version is a "fixed rate" where the interest rate remains, well, fixed. Another possibility, the "variable rate," is just that; the

© Springer Nature Switzerland AG 2019

D. G. Saari, *Mathematics of Finance*, Undergraduate Texts in Mathematics,
https://doi.org/10.1007/978-3-030-25443-8_9

interest rate changes. Here the value of the bond is $B(r, t)$ where $r$ is the interest rate, $t$ is the current date.

Bonds, as true with options and stock, are traded on the market. This means it is of value to determine what should be the current value of a bond $B_1(r, t)$. A first approach would be to mimic the Black–Scholes approach by hedging both within the bond market and between markets. Here is a problem: Earlier hedges had a $V(S, t) - \delta S$ form, but it is nonsense to try to hedge with

$$B_1(r, t) - \delta r. \tag{9.1}$$

How are you going to go short on 2% of interest? It resembles trying to go short on "February 2," which cannot be done in spite of fantasy movie's claim that this "Groundhog day" can repeat again and again.

A way to circumvent this puzzle is with a surrogate. In terms of bonds, find something else that is affected by the same interest rates. The choice can be almost anything, such as another bond, $B_2(r, t)$, with a different maturity date. With this choice, the hedging (the analysis to find how $B_1(r, t)$ is affected by changes in the $r$ value) is

$$f(r, t) = B_1(r, t) - \delta B_2(r, t). \tag{9.2}$$

Another complexity involves changes in the interest rate. While it is reasonable to speculate that

$$\Delta r = \mu r \Delta t + \sigma r \Delta X, \quad \Delta X \sim N(0, \Delta t),$$

a moment's reflection shows that this is silly. Why should the interest rate, or its random fluctuations, be proportional to the current $r$ value? Estimates on current changes are based on careful research that is not described here. Instead, assume that the change in the interest is given by

$$\Delta r = g(r, t)\Delta t + h(r, t)\Delta X, \quad \Delta X \sim N(0, \Delta t). \tag{9.3}$$

With this choice, Itô's Lemma becomes (and this is an exercise)

$$\Delta f = \frac{\partial f}{\partial r}\Delta r + \frac{\partial f}{\partial t}\Delta t + \frac{h^2(r, t)}{2}\frac{\partial f^2}{\partial r^2}\Delta t. \tag{9.4}$$

### Returning to Bonds

With the help of Equation 9.4, the Equation 9.2 expression can be handled. Namely,

$$\Delta f = \left[\frac{\partial B_1}{\partial r} - \delta\frac{\partial B_2}{\partial r}\right]\Delta r + \left[\frac{\partial B_1}{\partial t} - \delta\frac{\partial B_2}{\partial t}\right]\Delta t + \frac{h^2(r, t)}{2}\left[\frac{\partial^2 B_1}{\partial r^2} - \delta\frac{\partial^2 B_2}{\partial r^2}\right] \tag{9.5}$$

A first step is to eliminate explicit risk. The only place in Equation 9.5 where unchecked randomness and risk arise is with the $\Delta X$ component of the $\Delta r$ term. Setting this component equal to zero, or

$$\delta = \frac{\partial B_1}{\partial r} \Big/ \frac{\partial B_2}{\partial r} \tag{9.6}$$

captures the hedge ratio between the two bonds, which can be viewed as between changes of the specified bond and a proxy for interest rates.

While Equation 9.6 suffices for many purposes, a natural next step is to determine how the value of the bond compares with the general market. This can be done by determining what would happen should the $B_1 - \delta B_2$ portfolio be invested in some other market measure. What measure? That is up to the reader; perhaps the $B_1 - \delta B_2$ portfolio could be invested in a bank that has a different interest rate, or in options, or ... Ah, what fun!!

## 9.2  Dividends, or Other Embellishments

Other issues with respect to the value of a Call or Put include the effect of added payments, such as dividends. (A dividend is money that a company pays stockholders as their share of company profits.) Although dividends can be paid in sums at specified times, the simpler case considered here is where dividends are paid on a continual basis.

To carry out the analysis, let $d^*S\Delta t$ represent the rate of the dividend being offered as calibrated with respect to the value of the stock. A notational comment is that the star in $d^*$ is included to ensure that $d^*S$ is viewed as a value—a number—rather than the differential of $S$.

Clearly, this added influx of money affects the $\Delta t$ coefficient in

$$\Delta S = \mu S \Delta t + \sigma S \Delta X.$$

For intuition as to why this is so, consider the change of value in any item, such as your bank account with interest determined by $\Delta M = r M \Delta t$. This expression must be modified if money is being added (perhaps grandmother Lillian is sending in extra money) or subtracted (such as money removed for payment of college loans). For our purposes, however, this term can be safely ignored. The reason is that when computing $\Delta_{market} \Pi = \Delta_{market}(C_E(S,t) - \delta S)$, the explicit $\Delta S$ term is dropped (e.g., Equation 5.5, Exercise 1, Chapter 5) when defining the appropriate hedge ratio choice of $\delta = \frac{\partial C_E(S,t)}{\partial S}$. Recall, this is done to avoid the explicit risk resulting from $\Delta X$.

This dividend term *cannot* be ignored when computing $\Delta_{market} \Pi = \Delta_{market}[C_E(S,t) - \delta S]$ because the dividend revenue is money coming in, so it must be included. This leads to $\Delta_{market}[C_E(S,t) - \delta S] - \delta d^*S$. To explain, the sign of $-\delta$ determines whether you are going long or short. By going long, $-\delta d^* S$ is dividend money to be enjoyed. By going short, $-\delta d^* S$ is dividend money that is lost. After determining the hedge ratio $\delta$ (to drop the $\Delta X$ term), we obtain

$$\Delta_{market}\Pi = \left[\frac{\partial C_E}{\partial t} + \frac{1}{2}\sigma^2 S^2 \frac{\partial^2 C_E}{\partial S^2} - d^* S \frac{\partial C_E}{\partial S}\right]\Delta t.$$

The hedge with the market, as captured by putting all of $\Pi$ into a bank, remains the same. Thus the new Black–Scholes Equation, which reflects continuous dividends (or any continual payment based on the value and number of assets), is

$$\frac{\partial C_E}{\partial t} + \frac{1}{2}\sigma^2 S^2 \frac{\partial^2 C_E}{\partial S^2} + (r - d^*)S\frac{\partial C_E}{\partial S} - rC_E = 0 \qquad (9.7)$$

This expression resembles the standard Black–Scholes Equation.

### 9.2.1  A New Problem

The difference between Equation 9.7 and the standard Black–Scholes equation is that in one place the $(r - d^*)$ coefficient replaces the $r$ term. A first reaction is to seek some mathematical trickery to convert Equation 9.7 into an expression where the Equations 6.4, 6.9 solutions apply. Doing so requires finding some approach that will convert the $-rC_E$ term of Equation 9.7 into $-(r - d^*)C_E$.

If this is to be done, somehow an extra coefficient needs to be inserted into Equation 9.7. One mathematical trickery approach is to exploit the properties of exponentials and the product rule by defining

$$C_E = e^{g(T-t)}C_E^* \qquad (9.8)$$

where $g$ is an unknown constant; its value will be selected to accomplish what is desired. Using the product rule with $C_E = e^{g(T-t)}C_E^*$ leads to

$$\frac{\partial C_E}{\partial t} = \frac{\partial e^{g(T-t)}C_E^*}{\partial t} = -ge^{g(T-t)}C_E^* + e^{g(T-t)}\frac{\partial C_E^*}{\partial t} = e^{g(T-t)}\left[-gC_E^* + \frac{\partial C_E^*}{\partial t}\right].$$

All other Equation 9.7 partials are with respect to $S$, not $t$, so they treat this $e^{g(T-t)}$ multiple as a constant that can be factored out of the partial derivative. Thus, only the first partial (thanks to the product rule) introduces an extra term *that includes the sought after extra coefficient g*. All terms have the same $e^{g(T-t)}$ multiple, so Equation 9.7 becomes

$$e^{g(T-t)} \left[ \frac{\partial C_E^*}{\partial t} + \frac{1}{2}\sigma^2 \frac{\partial^2 C_E^*}{\partial S^2} + (r - d^*)S\frac{\partial C_E^*}{\partial S} - (r + g)C_E^* \right] = 0.$$

After dividing both sides of this equation by $e^{g(T-t)}$, what remain has the appearance of the Black–Scholes equation *if* $g = -d^*$. So, define $g = -d^*$, which means that Equation 9.7 becomes

$$\frac{\partial C_E^*}{\partial t} + \frac{1}{2}\sigma^2 S^2 \frac{\partial^2 C_E^*}{\partial S^2} + (r - d^*)S\frac{\partial C_E^*}{\partial S} - (r - d^*)C_E^* = 0 \qquad (9.9)$$

where the solution can be written down immediately by replacing the original interest rate $r$ in the $d_1$ and $d_2$ expressions with "pretend" interest rate $(r - d^*)$.

Namely,

$$C_E^*(S, t) = SN(d_1^*) - Ee^{-(r-d^*)(T-t)}N(d_2^*) \qquad (9.10)$$

where

$$d_1^* = \frac{\ln(\frac{S}{Ee^{-(r-d^*)(T-t)}})}{\sigma\sqrt{T-t}} + \frac{1}{2}\sigma\sqrt{T-t}, \quad d_2^* = \frac{\ln(\frac{S}{Ee^{-(r-d^*)(T-t)}})}{\sigma\sqrt{T-t}} - \frac{1}{2}\sigma\sqrt{T-t}.$$
$$(9.11)$$

### 9.2.2 Here Comes the Solution

The solution for a Call with dividends is given by the above along with Equation 9.8 to be

$$C_E(S, t) = \{e^{-d^*(T-t)}N(d_1^*)\}S - Ee^{-r(T-t)}N(d_2^*), \qquad (9.12)$$

where $d_1^*, d_2^*$ are defined above.

An interesting feature is associated with the American Call. Namely, when sketching the European Call (Section 8.1.1), it was discovered that $C_E(S, t)$, without embellishments, never crosses the $y = \max(S - E, 0)$ graph. In part, this is because the $N(d_1)$ coefficient for $S$ approached unity for large $S$ values. This feature meant there were no arbitrage opportunities with a European Call suddenly becoming American.

But the $S$ coefficient for Equation 9.12 approaches $e^{-d^*(T-t)} < 1$ for large $S$ values. The fact this slope is less than unity means that, eventually, the graph of Equation 9.12 *must* cross the $y = \max(S - E, 0)$ graph. When this is so, it unleashes arbitrage opportunities with the American option. In turn, this means that the graph of the American option follows that of the European option (Equation 9.12) until the graph hits the $y = \max(S - E, 0)$ line, when it then follows the $S - E$ line.

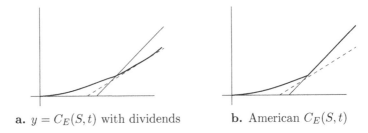

**a.** $y = C_E(S, t)$ with dividends          **b.** American $C_E(S, t)$

**Fig. 9.1** Dividends

(Arbitrage arguments of anticipation can be developed to assert that the curve does not have the Figure 9.1b kink, but that subtlety is ignored here.)

## 9.3  Numerical Integration

Another topic is based on the reality that, more often than not, finding an analytical solution for a partial differential equation is not possible. A substitute is a numerical solution. While a full course in numerical integration is required to fully appreciate what can be done, the spirit of the basic notions follows. A purpose of the following (woefully incomplete) description is to reflect the power of Taylor series and to hopefully encourage the reader to explore the fascinating topic of numerical computations.

### 9.3.1  The Heat Equation

Suppose the problem is to find a solution for

$$\frac{\partial u(x, t)}{\partial t} = \frac{\partial^2 u}{\partial x^2} \tag{9.13}$$

with given boundary conditions. To invent conditions, let $u(x, 0) = x^2$ for $0 \leq x \leq 1$, $u(0, t) = 0$, and $u(1, t) = 100$.

Stated in words, the temperature along the bar at time zero is specified by $u(x, 0) = x^2$. For all future time, a block of ice is placed on one end of the bar to create the boundary condition $u(0, t) = 0$ Celsius, while at the bar's other end, boiling water is placed to create the $u(1, t) = 100$ condition. The meaning of these boundary conditions is not important for our purposes, but they are required to solve the problem.

The idea is to replace Equation 9.13 with a discrete approximation by using the definition of a derivative and Taylor Series. Namely,

$$\frac{\partial u(x,t)}{\partial t} = \frac{u(x,t+k) - u(x,t)}{k} + o(k) \tag{9.14}$$

where the $o(k)$ expression represents error terms much smaller than $k$. This just means that the smaller the value of $k$, the more accurate the approximation obtained after dropping $o(k)$.

To find a representation for $\frac{\partial^2 u}{\partial x^2}$, use Taylor series. Here, there are the two expressions

$$u(x+h,t) - u(x,t) = \frac{\partial u(x,t)}{\partial x}h + \frac{1}{2}\frac{\partial^2 u(x,t)}{\partial x^2}h^2 + o(h^2) \tag{9.15}$$

and (replacing $h$ with $-h$)

$$u(x-h,t) - u(x,t) = \frac{\partial u(x,t)}{\partial x}(-h) + \frac{1}{2}\frac{\partial^2 u(x,t)}{\partial x^2}(-h)^2 + o(h^2). \tag{9.16}$$

Adding these two equations (which nicely cancels the $\frac{\partial u}{\partial x}$ terms!) and dividing by $h^2$ yields

$$\frac{\partial^2 u(x,t)}{\partial x^2} = \frac{1}{h^2}[u(x+h,t) - 2u(x,t) + u(x-h,t)] + o(1). \tag{9.17}$$

(so the $o(1)$ portion represents terms that go to zero as $h$ goes to zero).

Using Equations 9.14, 9.17, Equation 9.13 can be approximated by

$$\frac{u(x,t+k) - u(x,t)}{k} = \frac{1}{h^2}[u(x+h,t) - 2u(x,t) + u(x-h,t)] + o(1) + o(k). \tag{9.18}$$

Remember, the purpose of solving Equation 9.13 is to determine the behavior of $u$ in the future. This goal suggests rewriting Equation 9.18 to emphasize future behavior; solve for $u(x, t+k)$ to provide the value of $u$ at the future time $t+k$ based on what is happening now at time $t$.

Letting $\beta = \frac{k}{h^2}$ and, for the moment, ignoring the error terms (the o(k) etc terms), the desired expression is

$$u(x,t+k) \approx u(x,t) + \beta[u(x+h,t) - 2u(x,t) + u(x-h,t)] \tag{9.19}$$

where the error is $o(k)$. This means that the value of $u(x, t+k)$ at the future time is dictated by the current $u(x,t)$ value as modified by

$$\beta[u(x+h,t) - 2u(x,t) + u(x-h,t)].$$

For Equation 9.19 to yield reasonable answers, $h$, $\beta$, and $k$ must have small values (required to ignore the error terms)—the smaller their values, the more accurate the solution.

Equation 9.19 specifies future behavior at $u(x, t + k)$ by what happens now at $x$ (given by $u(x, t)$) as modified by the $\beta[u(x + h, t) - 2u(x, t) + u(x - h, t)]$ term. Of delight is how this expression offers intuition about how the $\frac{\partial^2 u}{\partial x^2}$ term affects the solution. This is because the Taylor series representation for $\frac{\partial^2 u}{\partial x^2}$ introduces the $\beta[u(x + h, t) - 2u(x, t) + u(x - h, t)]$ modifying expression, which shows how this term mixes up and combines neighboring behavior! And so, by using the information specifying the value of $u(x, 0)$ with Equation 9.19, the equation describes what happens at $u(x, k)$ for all values of $x$. After deriving this information, it can be used to find the value of $u(x, k + k)$ for the different values of $x$. The next step is to continue with other $x$ values.

To compute answers create a grid, as in Figure 9.2, where the vertical lines are $k$ units apart and the horizontal lines are $h$ units apart. All of the values (from boundary conditions) along the extreme vertical line to the left, which represent $t = 0$ are given by $u(x, 0) = x^2$. The goal is to find the values of $u(x, k)$, which are the values along the next vertical line.

As an illustration, notice the star ($\star$) on the line $t = k$. To find $u$'s value at this point by using Equation 9.19, substitute the given values of $u(x, 0)$ at the three bullet points. For instance, suppose $k = 0.01$, $h = 0.2$, so $\beta = \frac{k}{h^2} = \frac{0.01}{0.04} = \frac{1}{4}$. To find the $u(\frac{1}{2}, 0.01)$ value, substitute the appropriate $u(x, 0)$ values into Equation 9.19 to have

$$u(\tfrac{1}{2}, 0.01) \approx u(\tfrac{1}{2}, 0) + \beta[u(x + h, t) - 2u(x, t) + u(x - h, t)]$$
$$= (0.5)^2 + \tfrac{1}{4}[(0.52)^2 - 2(0.5)^2 + (0.48)^2].$$

To compute the $u$ value of the point immediately below the star, move the three bullets down one level and use them as above. In this manner, all $u$ values along the $t = k$ vertical line can be obtained. (There are some problems along the horizontal boundaries; just "double up" on them.)

**Fig. 9.2** A numerical approximation

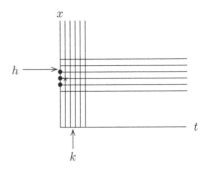

Once the values along the $t = k$ vertical line are found, use the same approach to find the values along the second, $t = 2k$, vertical line. In this manner, the solution can be approximated. (This is something that can be programed on an Excel program.)

## 9.4 What Is Next?

These notes provide an intuitive introduction for financial issues that are based on Puts, Calls, bonds, and so forth. Objectives included introducing tools and a way of thinking. As indicated in this chapter, once these tools are understood, they extend to a variety of new settings.

There is so much more to explore and be done. Throughout, attention was paid to the basic assumptions that are needed to create the tools and results. Most of these assumptions are reasonable—but primarily in quiet nicely behaved settings. This comment identifies one area which needs more research. From a mathematical perspective, for instance, those assumptions of independence, identically distributed, and so forth should be replaced with more realistic conditions. The assumptions of a constant $\mu$ and $\sigma$ need to be replaced with more general representations.

Much of this has been done, which means that these notes serve as an intuitive introduction. To go further, the reader is encouraged to improve her/his abilities in statistics, computer science, and almost all areas of mathematics.

There are many excellent books on options. For instance, although it was published a decade and half before the current millennium, the book by Cox and Rubinstein [2] has several insightful arguments that are worth reviewing. A standard choice is the textbook by Hull [6]; also check the references.

## 9.5 Exercises

1. Fill in the details for Equation 9.4.
2. Derive Equation 9.6.
3. After the $\delta$ value is determined (Equation 9.6), suppose Equation 9.5 is compared with the market as given by investing it in a bank that has a different interest rate of $r^*$. Find the associated Black–Scholes Equation.
4. Fill in the details to derive Equation 9.7.
5. Show how to obtain Equation 9.10.
6. Carry out the argument and details of the graph for an American Call with continuous dividends.
7. Use Equation 9.12 and the Put–Call Parity Equation to find the solution for $P_E(S, t)$.
8. Find the Equation 9.19 version for $\frac{\partial u(x,t)}{\partial t} = 2\frac{\partial^2 u}{\partial x^2}$.

9. Do the same for $\frac{\partial u(x,t)}{\partial t} = \frac{\partial^2 u}{\partial x^2} + \frac{\partial u}{\partial x}$. There are many answers for this question depending on how you represent $\frac{\partial u}{\partial x}$. One way is to use the Equation 9.17 approach of Equations 9.15, 9.16, but now use only the first derivative terms. Doing so leads to

$$\frac{\partial u(x,t)}{\partial x} = \frac{u(x+h,t) - u(x-h,t)}{2h}.$$

The rest of this problem should now be immediate.

10. Now try $\frac{\partial u(x,t)}{\partial t} = 3\frac{\partial^2 u}{\partial x^2} + \frac{\partial u}{\partial x} - u$.

11. Derive a version of Equation 9.19 for

$$\frac{\partial C}{\partial \tau} = \frac{\sigma^2}{2}\frac{\partial^2 C}{\partial x^2} + (r - \frac{\sigma^2}{2})\frac{\partial C}{\partial x} - rC. \tag{9.20}$$

Notice, Equation 9.20 is precisely the Black–Scholes Equation after using the change of variables $t = -\tau$ and eliminating the variable coefficients. While the heat equation may not be of interest to many readers, Equation 9.20 most surely is. Everything above applies to deriving numerical answers.

12. For the above system of the heat equation, find the value of $u(0.52, 0.01)$.

13. For the same boundary conditions and the above derived approximation (Problem 2) for $\frac{\partial u(x,t)}{\partial t} = \frac{\partial^2 u}{\partial x^2} + \frac{\partial u}{\partial x}$, find the value of $u(\frac{1}{2}, 0.01)$.

# References

1. Black, F., and M. Scholes. 1973. The Pricing of Options and Corporate Liabilities. *Journal of Political Economy* 81: 637–654.
2. Cox, J., and M. Rubinstein. 1985. *Options Markets*. New Jersey: Prentice-Hall.
3. Doerr, C., N. Blenn, and T. Van Mieghem. 2013. Lognormal Infection Times of Online Information Spread. *PLoS ONE* 8(5): e64349.
4. Golden, L., M. Wang, and C. Yang. 2007. Handling Weather Related Risks Through the Financial Markets: Considerations of Credit Risk, Basis Risk, and Hedging. *Journal of Risk & Insurance,* 74: 319–346.
5. Grönholm, T., and A. Annila. 2007. Natural Distribution. *Mathematical Biosciences* 210: 659–667.
6. Hull, J. 2018. *Options, Futures, and Other Derivatives*. Tenth edition. London: Pearson Education.
7. Jewson, S., A. Brix, and C. Ziehmann. 2005. *Weather Derivatives Valuation: The Meteorological, Statistical, Financial and Mathematical Foundations*. Cambridge: Cambridge University Press.
8. Lessler, J., N.G. Reich, R. Brookmeyer, T.M. Perl, K.E. Nelson, D.A. Cummings. 2009. Incubation Periods of Acute Respiratory Viral Infections: A Systematic Review. *The Lancet Infectious Diseases* 9(5): 291–300. https://doi.org/10.1016/S1473-3099(09)70069-6.
9. Petravic, J., P. Ellenberg, M.-L. Chan, G. Paukovics, R. Smyth, J. Mak, and M. Davenport. 2013. Intracellular Dynamics of HIV Infection. *Journal of Virology* 88(2): 1113–1124. https://doi.org/10.1128/JVI.02038-13.
10. Preston, F. 1948. Commonness, and Rarity, of Species. *Ecology* 29: 254–283.

# Index

© Springer Nature Switzerland AG 2019
D. G. Saari, *Mathematics of Finance*, Undergraduate Texts in Mathematics,
https://doi.org/10.1007/978-3-030-25443-8